Managing Geographic Information Systems

Managing

Geographic Information Systems

Second Edition

Nancy J. Obermeyer and Jeffrey K. Pinto

THE GUILFORD PRESS
New York   London

© 2008 The Guilford Press
A Division of Guilford Publications, Inc.
72 Spring Street, New York, NY 10012
www.guilford.com

Printed in the United States of America

This book is printed on acid-free paper.

Last digit is print number: 9 8 7 6 5 4 3 2 1

Library of Congress Cataloging-in-Publication Data

Obermeyer, Nancy J., 1955–
  Managing geographic information systems / by Nancy J. Obermeyer,
Jeffrey K. Pinto.—2nd ed.
    p. cm.
  Includes bibliographical references and index.
  ISBN-13: 978-1-59385-635-9 (hardcover)
  ISBN-10: 1-59385-635-0 (hardcover)
  1. Geographic information systems. I. Pinto, Jeffrey K. II. Title.
  G70.212.O24 2008
  910.285—dc22

                                                                2007031574

Contents

Chapter 1

The Continuing Need
for a Management Focus in GIS

It has been more than 10 years since the publication of the first edition of
Managing Geographic Information Systems. This chapter briefly outlines
the changes in geographic information systems (GIS) technology and in
the field more generally and makes the case that there is a need for this
extensively updated and enlarged second edition. After presenting our justification for this updated edition, we lay out the book, chapter by chapter.

As we complete this second edition of *Managing Geographic Information Systems*, the technology and its implementation have evolved and
changed dramatically. The technology itself has become increasingly easier to use, with the expansion of graphic user interfaces that make it ever
more accessible to nonexperts (so-called thin users). Accordingly, the
implementation of GIS has grown by leaps and bounds in terms of both
the number of users and the breadth of applications. This represents a
dramatic expansion of the technology's market penetration in the decade
or so that has passed since the publication of the first edition.

GIS and its cognate technologies—especially global positioning systems (GPS)—have become so commonplace that GIS has played a supporting role in a television series (*The District*), hand-held GPS devices are on
sale at discount department stores and offered as an option on many auto-
mobiles, and a radio-frequency identification device (RFID) has been
implanted in the arm of Mexico's attorney general as a demonstration

(*www.msnbc.msn.com/id/5439055/*). Dogs and other household pets are routinely fitted with a microchip that reveals the identify of beloved missing "Spike" or "Fluffy" and his or her owners, thus paving the way to a safe return home. (In spite of this technology, "Vivi," the microchipped Westminster Kennel Club dog show participant has not been reconnected with her owners more than a year after the whippet escaped from her kennel at the airport.)

But wait, there's more. Today consumers willingly provide an array of identifying information to retail establishments (both brick-and-mortar and online establishments) in exchange for special bargains, promotions, and other bonuses that are not available to anonymous shoppers. In return, the retailers that offer these bargains gain a great deal of information about each of their loyal customers along with the building blocks of a database that can help them guide their future business development activities. For the online consumer, the monitoring of their shopping habits usually generates a list of "suggestions" regarding future purchases based on past purchases, to which any regular customer of Amazon.com or Netflix will testify (coauthor Nancy Obermeyer included). And if that weren't enough, closed-circuit TV records our activities whenever we are within camera shot—which is whenever we are in most brick-and-mortar establishments (both public and private) and in some jurisdictions when we are in any public space, including on the roads and streets.

Many of us are aware of the indelible tracks we leave in the wake of our purchases. What some people may not know is how readily visible many of our tracks are to people who do not know us personally. Many local governments, for example, make tax records available online, permitting anyone with an Internet connection to learn more about us than we know ourselves. Some of these online databases, for example, the City of Milwaukee's, are available within the framework of an online, searchable GIS database. In another example, the tax records of property owners in Vigo County, Indiana, are available through an online search that provides names and addresses along with tax information (including whether or not the homeowner has paid his or her tax bill); this data set was finally attached to a base map in late 2006. In fact, the increased integration of GIS and its components with the Internet is another profound change for the technology, its users, and its managers.

These changes have had a profound impact on GIS and its management. Whereas the first edition of *Managing Geographic Information Systems* focused on efforts to bring the technology to organizations that had not yet implemented them—at that time, this included most organizations—this second edition explores many issues that were barely on the radar screen back in the day.

Purpose and Objectives

The purpose of this chapter and the book as a whole is to introduce the challenges that organizations face in managing their use of what has become a mature technology, one that has a tremendous capacity to affect the activities and productivity of a public or private organization. This book is intended to provide a combined theoretical and practical foundation for the effective development and use of GIS within an organization.

GIS has become a common tool in organizations within both the public and the private sectors. Increasing capabilities, decreasing costs, and easier-to-use interfaces have all contributed to the diffusion of GIS. As Goodchild (2005: 4) points out, "We are moving rapidly from a *concert pianist* model of GIS as a tool confined to experts, to a *child of ten* model in which the power of GIS is available to all, the obvious concerns about powerful and complex technology in the hands of naive users notwithstanding."

Managing GIS remains a two-pronged problem: mastery of the technology itself and understanding how to manage its effective use within an organization in the context of a specific institutional mission in service of a particular clientele. While there is far more literature available on GIS management today, this book aspires to address the primary issues associated with managing GIS technology and databases in an integrated and cohesive format, essentially providing "one-stop shopping" for its readers. This one-stop shopping, however, is designed to foster an increased interest in the individual chapter topics while providing bibliographic references that will lead the reader to more specific sources on topics of special interest.

The spread of GIS to a wider user base increases the importance of knowledge about managing this particular technology. Geography remains a key element of the technology, but meanwhile geographic knowledge remains less than ubiquitous. Moreover, the concerns raised by the use of the technology have expanded in both number and complexity. Today's GIS manager must be alert to issues that were barely articulated a decade ago. The chapter topics are designed to address this need.

The Chapters

Those of you who are familiar with the first edition will recognize a few updated chapters from that version, but you will also notice major changes and additional materials. These changes include an expansion from 11 to 15 chapters. This obviously includes topics that barely registered in the GIS community a decade ago, but it also includes extensive coverage and updating of chapters that have remained from the first edition. We describe the content of all the chapters below.

Chapter 1. The Continuing Need for a Management Focus in GIS

This chapter sets the stage for the need for this second edition. In addition, we make the argument that the key to managing an efficient and effective GIS remains, at bottom, a human challenge born of the need to understand how and why people are affected by and in turn affect GIS dissemination and use. We briefly describe the changing scene within which GIS exists and within which managers must operate, and continue by introducing the rest of the chapters.

Chapter 2. Geographic Information Science: The Evolution of a Profession

What began in the 1960s as a useful technology for managing data with a geographic location has evolved to become something far greater. GIS has become a full-fledged profession. This chapter begins with a theoretical discussion of the characteristics of a profession and then describes how GIS (or geographic information science) has developed these characteristics and therefore qualifies as a profession. One of the key advantages of this evolution is that the field has become more well defined, with a better articulated body of knowledge and clearer norms and conventions of conduct. These are all discussed in Chapter 2.

Chapter 3. The Role of Geographic Information within an Organization's Information System

In order to best understand the implications and use of geographic information, it is necessary to place the GIS within the context of a larger, fully integrated information technology (IT) system that provides managers with relevant information for performing their duties. The chief purpose behind an IT system is to aid managerial decision making by providing organizational members with comprehensive, comprehensible, and immediate information. This chapter elaborates the evolution of IT and then demonstrate how GIS fits into a comprehensive organization IT.

Chapter 4. Keeping the *G* in GIS: Why Geography Still Matters

This chapter is a holdover from the first edition, serving as a reminder of why geography still matters to GIS. GIS are a departure from the typical policy development tools because of their explicitly geographic component. One of the keys to unlocking the potential of GIS—and even more importantly, to avoid making serious mistakes—is a solid understanding of geography among GIS users. This chapter sheds light on geographic and cartographic principles that underlie GIS technology using examples from public policy analysis and cartography. The objective of this chapter is to raise the geographic consciousness of GIS users.

Chapter 5. GIS and the Strategic Planning Process

This chapter offers strategic decision makers and organization policy developers an understanding of how geographic information can be integrated into an organization's overall strategic planning process. We define the concept of strategic planning. We propose a general model of strategic planning that will serve as the basis for gaining a better understanding of all relevant elements in creating an organization's strategy. Finally, we analyze the role of GIS in developing comprehensive strategic plans and suggest that the type of information provided by a GIS makes it uniquely capable of enhancing the planning process for public and private organizations.

Chapter 6. Implementing a GIS: Theories and Practice

One of the key challenges in managing a GIS lies in gaining successful implementation of the technology in an organization. Although there are a number of impediments to its successful introduction, there are also several means by which an organization can better ensure its implementation. One critical factor in determining whether or not a GIS is likely to be accepted and used is the existence of an identifiable champion within the organization. These project champions have a tremendous impact on acceptance and use of new technologies. This chapter highlights the roles that champions play, the ways in which champions can impact the GIS, and some means by which organizations can begin to identify and make use of champions as they seek to gain widespread support for and use of their GIS.

Chapter 7. Organizational Politics and GIS Implementation

"Politics" is a term that conjures up a variety of images, most of them unpleasant. However, both research and practice demonstrate that organizational politics is really another term of the use of informal means of power and influence to help implement GIS. This chapter reviews the evidence for the importance of political behavior in implementing information and GIS technologies; offers logical propositions as to why politics occurs; establishes the normative, or positive, perspective on the use of political behaviors; and presents the findings from two GIS implementation cases that demonstrate the critical role politics can play in either promoting or derailing GIS implementation efforts.

Chapter 8. Economic Justification for GIS Implementation

One of the routine tasks associated with implementing a GIS within an organization is developing a cost–benefit analysis in order to justify the costs of the technology. This chapter takes a step-by-step approach to describing how this is accomplished, covering the basics, including the time-value of money. While addressing issues associated with tangible costs and benefits, the chapter also addresses intangible costs and benefits.

Chapter 9. Sharing Geographic Information across Organizational Boundaries

An intriguing dynamic that is currently being observed is the use of data sharing across organizational borders. This so-called interorganizational data sharing occurs for a variety of reasons, some of them economic (no one organization can afford to be the sole collector and storehouse for geographic data) and some of them based on efficiency (the need to pool resources among multiple organizations all needing the same data). This chapter takes an in-depth look at the data-sharing process, identifying the principal reasons (motivations) and means (mechanisms) by which organizations are willing to engage in sharing their geographic data with each other.

Chapter 10. Metadata for Geographic Information

Data are a crucial part of every GIS. This chapter focuses on the role of metadata in identifying appropriate data sets for use, as well as in sharing data with other organizations. The chapter goes into detail on the requirements for GIS metadata as developed by the GIS community with the framework of the Federal Geographic Data Center and encourages organizations to follow the recommendations to the best of their ability.

Chapter 11. Policy Conflicts and the Role of GIS: Public Participation GIS

In the first edition, we presented a hypothesis about the expanded use of GIS, suggesting that organizations would harness the technology to raise issues in the public arena. This chapter reasserts that original material in light of the development of what has come to be called "public participation GIS" (PPGIS), or sometimes "participatory GIS." Specifically, PPGIS is an application of GIS usually among nongovernmental organizations (NGOs) that brings local knowledge to a debate regarding a policy decision that affects local people. This has been a key and growing area of GIS implementation.

Chapter 12. Ensuring the Qualifications of GIS Professionals

One of the concerns among organizations implementing GIS is staffing. As GIS has become more common, so has the need either to evaluate and hire individuals or to train existing staff to work with the GIS. Chapter 12 explores this issue through a discussion of the debate on certification of GIS professionals, an idea that has become a reality in recent years. In particular, the chapter discusses the specific standards, in terms of education, experience, and active engagement with the GIS community, that GIS practitioners should have in order to develop and maintain their expertise in GIS. The chapter also discusses the growing importance of ethical behavior among GIS professionals.

Chapter 13. Legal Issues in GIS

The growth and diffusion of GIS technology has resulted in an expansion of the legal issues associated with it. When the first edition was published, discussion of legal issues was mostly found in disparate articles on the topic. Today the body of knowledge concerning legal issues in GIS has become more consolidated and cohesive. Because of this, and because of its growing importance to GIS managers, we include an extensive discussion of the most pressing legal issues in GIS in this revised edition.

Chapter 14. Ethics for the Professional GIS

This chapter discusses the rapidly evolving topic of ethics among GIS professionals. Early discussions of GIS ethics occurred in the beginning of the 1990s, but it has taken the development of certification of GIS professionals to bring this important matter from the talk forum to the action forum. Today, there is a GIS code of ethics and procedures to encourage GIS practitioners to abide by this code are nearing completion. This chapter discusses a topic that is of concern to everyone who uses a GIS.

Chapter 15. Envisioning a Future

The final chapter provides a brief summary of the key points of the book. More importantly, it suggests future directions in GIS that will influence the evolution of management issues, and discusses their implications.

Conclusions

The revised edition of *Managing Geographic Information Systems* represents our efforts to offer a challenge to the community of GIS practitioners as they manage their systems in an ever-changing environment. As the technology continues to proliferate and mutate, more and more individuals will find a need to understand not only how to run the software of their

GIS, but how to make best use of the technology within their specific organizational setting and in compliance with the best practice of the field.

We have done our best to be reasonably comprehensive in our coverage of topics, but because the field is shifting quickly, it is a moving target. For example, although we allude to the important role of the Internet in GIS, we do not include a chapter on this topic specifically at this time. Still, we hope you find value in our current effort, and we welcome your comments on our work.

Chapter 2

Geographic Information Science
EVOLUTION OF A PROFESSION

Ten years ago, the GIS community was well on its way to becoming a profession. At that time, it possessed most of the attributes of a mature profession and seemed well on its way to developing those that remained. Today, GIS has evolved into a mature profession, based on its possession of several key characteristics identified by Weber (1946) and Pugh (1989), including a body of knowledge, a social ideal, a professional culture, and a code of ethics (among others) Current initiatives among GIS professionals include continuing the development of a body of the knowledge in the field and making efforts to encourage ethical behavior through education and adoption of a code of ethics. This chapter discusses two changes in the GIS community: the use of the term "geographic information science" and the development of GIS as a profession.

From GI Systems to GI Science

The evolution of the term the "geographic information science" to describe the field of GIS is one of the many developments of the 1990s. As Goodchild (2005: 1) recounts, in 1990, the president of the Association of American Geographers described GIS as "nonintellectual expertise." Both Goodchild and his codirector of the National Center for Geographic Information and Analysis (NCGIA) David Simonett recognized the need for a "strong emphasis on science and theory." It was Goodchild who subsequently coined the phrase "geographic information science."

In taking this important step, Goodchild also defined the term as "research on the generic issues that surround the use of GIS technology, impede its successful implementation, or emerge from an understanding of its potential capabilities" (Goodchild, 2005). Mark (2003) lists and describes the components of geographic information science.

The first element of geographic information science is ontology and representation. This includes an examination of the concepts used within the field. This idea has been expanded to include the concepts as used by different groups that use GIS. Data modeling and representation are part of this element (Mark, 2003).

The second element of geographic information science is computation. This element begins with qualitative data reasoning and computational geometry. It also includes efficient indexing, retrieval, and search in geographic databases, as well as spatial statistics and other geocomputation topics. Cognition is the third element of geographic information science. It includes cognitive models of geographic phenomena, and human interaction with geographic information and technology (Mark, 2003).

Another critical element of geographic information science is applications, institutions, and society. There are several parts of this category: acquisition of geographic data, quality of geographic information, and spatial analysis. Of particular relevance to this book is the final part of this category: geographic information, institutions, and society (Mark, 2003).

The final two elements of GI science are time and scale, or what Mark (2003) describes as "cross-cutting research themes."

Management issues fall squarely under the "geographic information, institutions, and society" element of geographic information science. This book covers several research topics in this element, including economic and legal aspects of geographic information, and changes in organizational efficiency, effectiveness, equity, and power in society (Mark, 2003). The professionalization of GIS is another topic within this category.

Professionalism in GIS

Webster's Dictionary defines *profession* as "a calling requiring specialized knowledge and often long and intensive academic preparation; a principal calling, vocation, or employment; [and] the whole body of persons engaged in a calling." This definition is consistent with notions of professions and professionalism embedded within Max Weber's theory of bureaucracy.

Weber and Professionalism

Writing at the turn of the 20th century, Weber (1968b) described bureaucracy as an eminently enduring organizational model (Gerth & Mills, 1976). A major reason behind the staying power of bureaucracy is professionalism. Professionalism and professions are based on *expertise*, that is, a specialized knowledge or skill, and the ability of the profession to protect its expertise from outsiders. This specialized knowledge or skill required to demonstrate expertise is unique to each field. Weber identifies expertise as a prerequisite to the development of a profession.

Expertise is inherently both field-specific and time-specific. For example, leeches were once a common and acceptable course of treatment within the medical community. Today, lasers, new imaging technologies, and laparoscopy are important medical tools over which modern physicians must develop mastery if they are to be considered experts. Similarly, in the field of GIS, paper and pens have been supplemented (and in some cases replaced) by computerized hardware and software. As innovations diffuse within a field, the specialized knowledge of that field shifts to include them, as well as to eliminate obsolete techniques and ideas.

Weber further notes the importance of developing and closely guarding from outsiders the body of knowledge or expertise that forms the foundation of the profession (Gerth & Mills, 1976: 233). Professions typically use education and sometimes certification examinations as a means to limit entry into the profession. In addition, professional publications and networks facilitate the development and diffusion of a common language—sometimes better described as "jargon." This shared language serves a valuable function: it helps to identify who is a member of the profession and who is not, and may be used deliberately to make entry into the profession more difficult.

While the development of expertise may serve positive purposes (e.g., setting a standard of competency), Weber (1968b) raises concerns about the elevation of technical experts to the status of a mandarin caste. He notes that many professions gain a virtual monopoly in their area of expertise, which makes it very difficult for outsiders to evaluate the performance of members of the profession. The medical profession is a prime example (Berlant, 1975).

Similarly, Habermas (1970) suggests that experts may use their specialized knowledge to build a technocracy, thus gaining hegemony within their profession. Likewise, Cayer and Weschler (1988: 45) note that the expertise of professions and their concomitant control over information may lead to a concentration of power within the profession. There is a thin line between the concentration of expertise necessary to assure competency within a field and the use of expertise to create a technocracy. Given the technical nature of GIS, as members of the GIS community we should be concerned about the potential for creation of a GIS technocracy within the field.

Pugh's Six Characteristics of a Profession

Pugh (1989) identifies six characteristics of a profession. Most of these characteristics are self-evident; others need some explanation. The first of these characteristics is a cast of mind or a self-awareness, an acknowledgment by the professional that he or she is a member of a distinct profession. For example, when someone identifies himself as a doctor or herself as a lawyer, they are expressing a professional self-awareness.

The second trait of a profession is the possession of a unique body of knowledge necessary for the performance of professional duties. The idea of a body of knowledge is consistent with Weber's notion of expertise.

As the profession coalesces, it develops a third trait, what Pugh (1989) calls a "a social ideal to unify those working within an occupation." As an example, Pugh suggests that "for public administration, the consolidating vision was a knowledgeable, responsible, and proficient public service, the humane and efficient promotion of the common defense and general welfare, and the promotion of democratic institutions" (2). Weber suggests that this "inner devotion to the task, and that alone, should lift the scientist to the height and dignity of the subject he pretends to serve" (quoted in Eisenstadt, 1968: 297). In some sense, the social ideal appeals to the highest goals and aspirations of members of a profession, in terms of both competence and expertise; it helps lay the foundation for the development of an ethical professional community (Obermeyer, in press). We may think of this characteristic as a professional culture.

Eventually, as the profession evolves, members of the professional community join together formally to create a professional organization, the fourth characteristic of a profession. Frequently, professional organizations establish one or more journals, newsletters, electronic mail networks, or a variety of other mechanisms for promoting communication among the members.

These publications and networks become integral means to continue the growth, development, and maintenance of the profession's expertise as members share new ideas and refine (and sometimes eliminate) old ones. In addition, these publications and networks facilitate the development of a professional jargon, which serves a valuable function by helping to identify who is a member and who is not. At times, jargon may be used deliberately to make entry into the profession more difficult.

The fifth trait of a profession is "a hall of fame, a gallery of luminaries" (Pugh, 1989: 3). Individuals become part of this hall of fame by performing works in support of the profession, including theoretical and scholarly contributions, teaching and mentoring activities, and general advocacy on behalf of the profession.

Finally, a mature profession has a code of ethics. A code of ethics implies that the profession not only takes responsibility for a standard of competency among practitioners, but it endeavors to assure that its members will use their expertise ethically at all times. Professions may adopt any of several mechanisms to encourage ethical practice, including peer pressure and sanctions such as fines, suspensions, or even expulsion from the profession.

A Unified Model of a Profession

By including the essential elements of a profession as described by Weber and Pugh and combining similar or overlapping characteristics, it is possible to identify five key elements of a profession (Obermeyer, 1992, 1994):

1. The existence and growth of a unique body of knowledge (expertise).
2. The rise of a professional organization.
3. The evolution of a shared language.
4. The development of a professional culture and lore (including a "hall of fame").
5. A code of ethics.

Using these criteria as the basis of evaluation, it is clear that GIS is a profession. The evidence is presented below.

The Evidence

There is ample evidence to suggest that a GIS profession has evolved. As we will show, it meets all five criteria fully.

Unique Body of Knowledge (Expertise)

Professional expertise can be found in two separate areas: research and teaching. Generally, expertise is maintained and shared through the written (or, more recently, the electronically transmitted) word. The GIS community has a growing body of expertise, both in research about GIS and in the teaching of GIS. In recent years, the University Consortium on Geographic Information Science (UCGIS) has played a leading role in developing and organizing this body of knowledge through its major initiatives on GIS Body of Knowledge and its Model Curricula (*www.ucgis.org*).

Much of the early research on GIS existed in gray or fugitive literature, such as proceedings of professional meetings. As GIS has evolved as a profession, the literature in the field has become easier to find because it is available in more mainstream sources. For examples, there is a growing list of texts and collected readings on GIS, beginning with books by Aronoff (1989), Burrough and McDonnell (1998), Huxhold (1991), and others. The GIS reference *Geographical Information Systems: Principles and Applications* (Longley, Goodchild, Maguire, & Rhind, 1991, 1999) is a prime example of and source of expertise in the field.

But today there are many more books on GIS, encompassing every aspect of the technology itself as well as its many uses. There are books on GIS applications that include environmental analysis, spatial modeling, use of GIS as a tool for empowerment, GIS and its application in transportation, and many, many more. The market for GIS books has grown dramatically, providing an incentive to publishers to produce books to feed this hungry market.

In addition, articles on GIS are increasingly available in a variety of scholarly journals in fields such as geography, urban planning, landscape architecture, and surveying. Moreover, the creation of the *International Journal of Geographical Information Systems, Transactions in GIS*, and the renaming of the journal *Cartography* to *Cartography and Geographic Information Science* have resulted in a spectacular growth in scholarly literature specifically devoted to increasing, maintaining, and sharing GIS expertise.

Professional Organization

Evidence that the GIS community is well organized can be found in the early success of the annual GIS/LIS conference in the United States and the European GIS Conference (EGIS) in Europe throughout the 1990s. In the United States, the now-defunct GIS/LIS was cosponsored by five separate organizations: the Association of American Geographers (AAG), the American Congress on Surveying and Mapping (ACSM), AM/FM International, the American Society for Photogrammetry and Remote Sensing (ASPRS), and the Urban and Regional Information Systems Association (URISA). GIS specialty groups exist within these and other organizations as well. Today, GIS organizations are plentiful and increasingly specialized, eliminating the need for the early collaborative GIS/LIS format.

One of the most significant professional organizations within the GIS community is the University Consortium for Geographic Information Science (UCGIS). UCGIS is an organization whose members are primarily institutions of higher education within the United States. In order to qualify for membership, universities must demonstrate an interdepartmental collaboration surrounding GIS. In addition to its role in developing expertise in GIS (as noted above), UCGIS also sponsors a variety of activities, including winter and summer meetings where members may exchange ideas and develop collaborative activities.

Whereas UCGIS functions at the level of an educational institution, another recently founded organization, the GIS Certification Institute (GISCI), is designed to foster professionalism among GIS practitioners (*www.gisci.org*). GISCI is a spinoff of the URISA, which gave it a home and incubated it in its early days, until it began operating independently in 2004. Growing out of a desire to foster competent and ethical behavior among GIS practitioners, GISCI has established a means by which individuals may demonstrate that their education, experience, and contribution to the GIS community meet a set of standards devised to identify them as GIS professionals. Furthermore, GISCI has established a code of ethics and procedures for addressing ethical violations by its members (see "Code of Ethics" below).

Shared Language ("Jargon")

The development of expertise in GIS, along with the coalescence of the GIS community as an effective, functioning group, has promoted the evolution of a shared language. The GIS community speaks a jargon unto itself.

For example, when we say "GIS," we mean "geographic information systems" (not, for example, "guidance information systems," which is a real computerized system used by high school counselors). When we mention "GBF/DIME," we understand that this was a system used by the U.S. Bureau of the Census and is a predecessor of the current "TIGER" files. However, members of the GIS community would never confuse the "TIGER" files with a large, orange-and-black striped member of the feline family.

Similarly, we readily throw around phrases such as "object-oriented," use acronyms like "DLG," and discuss a variety of proprietary GIS, including "ArcGIS," "Idrisi," "GRASS," "MapInfo," and others.

A discussion among members of the GIS community would probably make little sense to an outsider, both because of the technical nature of GIS and because of the shared language that the community has evolved and which its members use when speaking among themselves.

Professional Culture and Lore, Including a Hall of Fame

The notion that a profession develops its own culture and lore is central to the creation of a distinct professional image. In this context, the professional culture is expressed in terms of networks of GIS managers, practitioners, and scholars; the mentoring process that often exists within and across organizations (including universities); and the celebration of important milestones in the profession.

Members of the GIS community identify several important milestones in the development of geographic information systems. For example, Waldo Tobler's "Map In, Map Out" research in the late 1950s is regarded as a key first step toward the creation of digital spatial data. In 1964, Roger Tomlinson's development of the Canadian GIS, considered by many in the community to have been the first true GIS, is another major milestone. Similarly, the adoption by the U.S. Census Bureau of the GBF/DIME files (and later the TIGER files) represents another important watershed.

The idea of a "lore" also refers to the collection of myths, stories, and a hall of fame that includes and honors early pioneers in the field, whether in an unofficial way or in an official hall of fame. GIS users frequently speak among themselves of the value and benefits of GIS, firm in the belief that GIS can help improve decision making in both the public and the private sectors. An important mission within the community is promoting the adoption of GIS for a wide and growing variety of applications. In general, they have been quite successful in pursuing this mission, as shown by the proliferation of GIS in a growing variety of applications. Furthermore, much like the manner in which IBM employees revere the memory of Thomas Watson Sr. or Disney employees look back to the impact of Walt Disney on their company, the GIS field is in the process of developing its own hall of fame made up of some of the early researchers and technical giants in the field. People such as Waldo Tobler, Roger Tomlinson, Ron Abler, Jack Dangermond, Robert Aangeenbrug, Duane Marble, and others all figure prominently in the development and evolution of the GIS field into an identifiable profession.

Code of Ethics

As the GIS field has grown and expanded into new application areas, many teething problems emerged, as they are bound to do with the introduction of any new technology and the concurrent growth in the number of people who use it. As these changes have occurred, there has been an increase in the need to address some of the unforeseen side effects of the use of technology, as in the legal problems resulting from cases of rights of privacy versus expanded access to information. Pugh (1989) makes the point that it is usually at this stage that some dialogue on shared concerns begins to make itself heard as GIS professionals attempt to establish a set of rules of behavior, comprising some form of a code of ethics.

One of the more significant ongoing activities of the GIS Certification Institute is its development of a code of ethics, which GIS practitioners who become certified by GISCI must sign and by which they must abide. GISCI's code of ethics emphasizes both competency and ethical behavior. It requires that GISCI-certified GIS practitioners embrace the spirit of a code of conduct and comply with specific rules of conduct developed by the GIS Certification Institute (GISCI, 2007).

The implementation of geographic information systems is not a value-neutral endeavor, a point that is echoed by a growing chorus within the geographic community. Researchers such as Dobson (1993), Goodchild (1993), Pickles (1993), and Sheppard (1993) emphasize that a Pandora's Box of societal repercussions (both good and bad) are bound to accompany the widespread adoption of GIS, particularly in the public sector.

This brings us back to the point that the development of a code of ethics has been a necessity for the GIS community. As individual GIS practitioners, we must first accept the enormous capabilities that GIS brings and take individual responsibility for our own actions. However, it is also necessary for us to develop the parameters of ethical behavior for the profession in order to prevent, to the extent possible, the unethical or immoral behaviors of others. The GIS Certification Institute has begun the effort with its code of ethics and its procedures for addressing violations of the code.

GIS: A New Profession

Without a doubt, what existed as a "GIS community" when the first edition of *Managing GIS* was originally published can now justifiably be called a new "profession." There can be no doubt that the community has expertise, a shared language, as well as a professional culture and lore (including candidates for a GIS hall of fame), which all were present when the first edition of *Managing GIS* came out. Added to these elements are GIS organizations (UCGIS and GISCI) and a code of ethics.

Chapter 3

The Role of Geographic Information
within an Organization's IT

In this chapter we offer an introduction to the development and use of information technology (IT) within an organization. In order to understand the implications and use of geographic information, one needs to place the GIS within the context of a larger, fully integrated information system (IS) that provides managers with relevant information for performing their duties. We show that the purpose behind an IS is to aid managerial decision making by providing an organization's members with data and other forms of information that are comprehensive, comprehensible, and of immediate use. As a result, in this chapter we develop an overall picture of the importance of IS for modern organizations, the duties of management at and the information needs across various management levels, the nature of the classical decision-making process and the ways in which an integrated IS can affect managerial decision making, the manner in which information is gathered and processed, and the specific role that geographic information plays within the context of this larger framework.

The society within which we exist and operate has become—and will continue to become—increasingly complex and fast paced. Within the private sector, competition takes place at the international level in a number of industries. As a result of these external pressures, the cycle time for new product innovation has decreased rapidly in an effort to speed up time to market in order to meet consumer needs. The public sector is equally affected by the faster and faster pace of our society. Local, county, and federal governmental agencies are being called on to take an increasingly proactive role in the management of infrastructure, land use, natural resource development, surveying, and a host of other activities related to the more efficient management of our urban and rural environments and natural resources.

Information has become a valuable and often expensive resource in today's society. The rapid rise in the creation and expansion of IS departments within organizations lends credence to the importance that is attached to providing managers with timely and useful information to enable them to better perform their duties through more effective decision making. In order to make clear the role of information within the operations of organizations, we must define exactly what is meant by the term. *Information* is data that has been converted, or operationalized, into a meaningful and useful context. Once such a context has been agreed upon, the information is of considerable value to specific organizational members, who use this information in an effort to arrive at better (i.e., more effective or more efficient) decisions.

Owing to the business world's increased need for precise and useful information, one of the more recent phenomena in the field of organizations and management has been the rapid rise of IS development and use. When most of us visualize an IS, we typically think of it in terms of technology—for example, we envision the information infrastructure of hardware, software, data storage, and networking. In reality, however, it is often more appropriate to think of IS as actually comprising three distinct elements that work in collaboration: the information technology itself, people, and processes. To be effective, these three elements must work together in harmony. An IS, then, refers to a system of people, resources, and procedures that collects, transforms, and distributes information to relevant organization members. The organization's IT revolves around the actual technical devices, concepts, and tools used in the system (Pearlson, 2001). For an IS to be effective, it must supply managers with information that is rapid, comprehensive, and accurate. Furthermore, it is important to emphasize that the most important aspect of information provided by a system is its usefulness to its end users—that is, managers. While there may actually be many methods for collecting and disseminating information, for the purposes of this chapter we concentrate on the activities of computer-based ISs.

An Overview of Management

Above we noted that IS were developed as a tool to enable managers to better perform their jobs. Such an observation, however, begs the larger question of the role that managers are expected to play for organizational success—in other words, What is it that managers do? In order to help our readers to gain a sense of the effect of IS on the process of management, we next devote some attention to a discussion of exactly what constitutes management. In other words, what are some of the specific duties and activities that correspond to the role of a manager within a public or private organization? Literally thousands of books have been written over the last century on the process of management, how it works, what the specific duties are, and how to improve it. Of all the works on the study of management over this period of time, one of the most influential is *General and Industrial Administration* by Henri Fayol (1916/1949). In this book Fayol outlined his views on the proper management of organizations and of the people in them. He presented five primary roles of management: planning, organizing, supervising, staffing, and controlling. These categories, although over 75 years old, have formed the basis of almost all subsequent work in the field of management. Each of the five primary tasks are defined in the following paragraphs.

Planning

The role of *planning* requires the manager to develop a set of goals and objectives and to create both short- and long-term plans for achieving these goals. Long-range plans are often broad, general outlines of where a company or a specific department wants to be in 5, 10, or even 20 years. Short-term goals are established to address and focus attention on specific targets that the organization seeks. These targets are seen as contributing and complementary to the organization's efforts to achieve its long-term objectives. For example, at General Electric in the early 1980s, Jack Welch, the new chief executive officer of the organization, formulated his famous "One, or two, or out" rule. In other words, each operating division within General Electric would, within 2 years, either become number one or number two in its specific product industry or it would be sold. With this long-term objective as their backdrop, operating managers within each of GE's strategic business units were required to formulate short-term plans for gaining a commanding share of their individual markets.

Organizing

Organizing refers to the methods by which managers organize, or make sense of, the work environment. The standard methods used to organize include the development of an organizational structure and operating rules and procedures. To illustrate: As an organizing function, a manager may choose to change the nature of the reporting structure within his or her specific department, either increasing or decreasing the number of levels of management between the workers and him- or herself. Another example of the use of organizing through standard operating procedures could be the establishment of a rule that all purchases or other expenses in excess of $1,000 must be approved by the department manager.

Supervising

Above all else, managers need to recognize that their primary responsibility is that of human resource management. In other words, managers are essentially involved in the *supervising* role. Their success or failure hinges on their ability to lead, motivate, and develop their subordinates. As effective leaders and motivators, managers are required to provide their employees with both the opportunity and the means to be productive. Furthermore, within their supervisory function, managers are sometimes called upon to act in a guidance mode with their subordinates, offering counseling and support for those who need it.

Staffing

The process of *staffing* involves the selection and professional development of organization personnel. In essence, it refers to the manager's responsibility to ensure that the right person, with the right training, occupies the right job. Staffing activities can actually be quite varied, from performing interviewing and hiring duties to providing job- and skill-training opportunities for organization personnel.

Controlling

Controlling refers to a manager's duty to monitor the activities of his or her subordinates in order to ensure that all activities are performed effectively and, in cases where deviations from plans are noted, to provide the necessary corrections. As a result of the monitoring process, managers may find it necessary to modify either employee performance or their initial plans. That is to say, if the manager observes that subordinates are performing to their maximum and yet are unable to achieve the targeted objectives, it would then become necessary to modify the initial projections in order to bring the objectives more into line with reality. On the other hand, if employees are not performing up to their potential, the manager may have to provide additional training or even correction and discipline, if appropriate. Control is most often found in the form of feedback, whereby a manager receives a report on each subordinate's job performance and, on the basis of this information, determines whether or not some form of correction is needed.

The common factor underlying each of the five duties of management is the need to make timely, informed, and accurate decisions. In order to make the most efficacious decisions, managers need to receive the types and quantity of information that will enable them to best perform their jobs. It is with this purpose in mind that organizations have developed and introduced a variety of IS into their operations.

The Role of Information Systems

The primary purpose of IS is to provide managers with information that is complete, accurate, and timely and that will enable them to make decisions that are more efficient and effective. As noted by Hutchinson and Sawyer (1992: 471–472), an IS is created to satisfy a manager's need for information "that is more summarized and relevant to the specific decisions that need to be made than the information normally produced in an organization and is available soon enough to be of value in the decision-making process."

IS serve the needs of managers in two ways. First, they provide a sense-making function in that they assist management in understanding the complex nature of the relationship between the organization and its environment. By having access to needed information of a readily usable nature, managers are able to make more informed and, arguably, better decisions. A good IS aids in gathering data and processing internally useful *intelligence information* as well as externally disseminated *public information* (Schermerhorn, 1989). Within an organization intelligence information is the basis upon which key decision makers chart long-term objectives. Public information is derived from the environment and allows an organization to engage in a wide variety of public activities, including image building, advertising, political support, and so forth.

The second way in which an IS serves the needs of managers is through timeliness. Obviously, information that arrives late or incomplete is of almost no value. In order for managers to reap the advantages of an IS, the system must materially influence the *way* in which managers arrive at decisions as well as the *type* of decisions they make. Furthermore, managers need to be aware in their own minds that these new, information-assisted decisions are in some sense superior to the old method, either through time savings or the enhanced effectiveness of the decisions themselves. It is also important to note that there is little in common with the types of decisions made at different levels within an organization. As a result, an IS must provide a variety of differential pieces of information so that the information can be accessed and can assist in supporting the decisions made by managers at different levels in an organization.

The Role of Information within an Organization's Operations

One useful method for visualizing the operations of an organization relative to its information needs is to think in terms of a "process-oriented" model of its operations. Michael Porter (1985) of Harvard University has created a useful model for understanding such process operations, the *value chain model*. Figure 3.1 illustrates the logic underlying the value chain model by delineating the various components of an organization's operations, based on the goal of value creation. If we consider the organization's principal activities of inbound logistics, operations, outbound logistics, marketing and sales, and service as being sequentially linked, we can see the operational flow by which a private organization will transform its raw materials into a product or service having value. Other necessary elements of the value chain consist of the activities of support functions including infrastructure, human resources, technology development and innovation, and procurement. Porter's model demonstrates a process flow in which the organization has established a sequential, value-creating process to its operations.

The value chain model is particularly relevant to understanding the usefulness of an organization's IS because it is easy for us to observe the logical flow of operations and the likely types of information that will be needed at each stage in the process. When developing a strategic view of

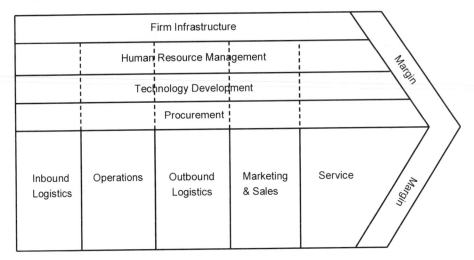

FIGURE 3.1. Porter's value chain model. Reprinted with the permission of The Free Press, a Division of Simon & Schuster Adult Publishing Group, from *Competitive Advantage: Creating and Sustaining Superior Performance* by Michael E. Porter. Copyright 1985, 1998 by Michael E. Porter. All rights reserved.

IS, we first need to anchor it to such a model that illuminates the demands not only for certain types of information but when they are most needed. For example, information related to the establishment of warehouses and distribution centers is key as an element of the outbound logistics of the organization. Therefore, in creating a strategic vision for the uses of information, we first need to understand the nature of the input–output flow of our operations, whether service-based or manufacturing, and then structure relevant information to be of the right nature and available at the points it is most needed.

Information Needs across Organization Levels

The first step in designing an IS is to determine what sorts of information management needs. This task is more difficult than it may at first appear, particularly as one moves upward within an organization's hierarchy. Top managers are often required to perform in a capacity that is characterized by greater ambiguity and that requires more generalized knowledge rather than specific functional expertise. Table 3.1 is intended to illustrate this point. It presents a summary of some of the different types of activities performed at various organization levels. Furthermore, it suggests the types of information that would be most useful to and appropriate for managers across these different organization levels. As you can see, first-level managers usually perform tasks that consist of implementing operational plans as developed by higher level management. For example, a first-level manager's duties may include scheduling production runs, assigning resources across various tasks, and transacting day-to-day business activities. Consequently, the types of information that first-level managers need is usually tied directly to the specific tasks they are called upon to supervise. Defect reports, exception reports, and adherence to budgets and schedules are examples of some of the concrete, specific types of information that are useful to first-level managers and that can be operationalized. Ideally, any diagnostic information that can help these individuals perform their duties more efficiently is valuable.

Middle-level supervisors also have a variety of duties, oftentimes of a more general and ambiguous nature than those assigned to first-level managers. Middle-level supervisors are usually called upon to find methods for implementing higher level strategies. As a result, they are tasked with the need to formulate operational plans and objectives that will allow

TABLE 3.1. Information Needs at Different Management Levels

Level	Activities	Information Needs
Top management	Strategy formulation: the establishment of long-term objectives and plans, making strategic decisions	Wide ranging: many sources are required, both internal and external to the organization; there are external opportunities and threats and internal strengths and weaknesses
Middle management	Formulation of plans for achieving strategic objectives: making a specific variety of operational decisions including resource allocation, employee evaluation, and short-term goal setting	Mostly internal: includes a combination of general and specific information requirements
First-level management	Performance of well-defined tasks: making short-term decisions, transacting day-to-day business	Diagnostic: designed to enable correction of deviations from specific schedules and budgets; best information is measurable

for the successful implementation of business strategies. Furthermore, they make operational decisions in support of these plans. For example, a middle-level manager who has been charged with increasing productivity in a series of midwestern plants may act to fulfill that requirement by reallocating human and financial resources to the midwestern region. As you can see, the type of information that a middle-level manager may require is more general and wide-ranging than what would be needed by a first-level supervisor. In addition to simple production reports, the middle-level supervisor in this example would also need financial and profitability data, as well as information relating to manpower and to budgetary slack.

Finally, top management operates in a very different manner from managers at other organization levels. Top managers engage in such activities as formulating long-term goals, making strategic decisions regarding corporate direction, determining products to be developed and produced, and securing a variety of scarce resources on which the company depends for survival. In order to most effectively engage in new strategy formulation and goal setting, top managers require a wealth of information that is not of interest to managers at lower levels. For example, many of the information needs of top management are external—that is, these needs require that an organization's IS analyze and provide data on general trends in the marketplace, on changes in governmental and economic policies, on consumer patterns and tendencies, and so forth. These types of information are in direct contrast to those needed by lower level managers. First-level managers are often provided with concrete, tangible information that enables them to compare actual progress to production or output plans and, where appropriate, to make necessary corrections. On the other hand, top management, which is engaged in a series of more ambiguous activities, requires a wealth of additional information from the external environment in order to chart the most effective courses for the organization in the future.

Another way to view the use of information at various organization levels is to recognize that not only is the information provided often tailored to suit the needs of individuals at different levels, but the information systems themselves may, in fact, be altered to more optimally address the needs of different managerial hierarchies. Consider the pyramid structure shown in Figure 3.2. The pyramid denotes four identifiable levels of managerial behavior and the major types of systems that can best provide the information these levels demand (Laudon & Laudon, 2001).

Type of Systems

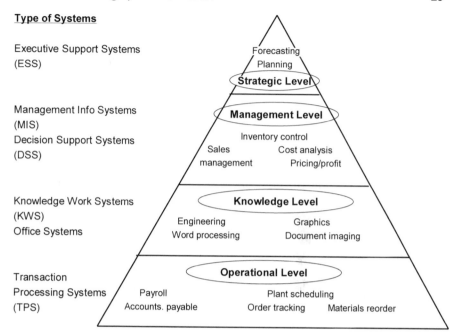

Executive Support Systems
(ESS)

Management Info Systems
(MIS)
Decision Support Systems
(DSS)

Knowledge Work Systems
(KWS)
Office Systems

Transaction
Processing Systems
(TPS)

FIGURE 3.2. Pyramid hierarchy of IS.

At the top of the pyramid are the strategic-level systems, usually defined in terms of the need to provide broader strategic information aimed at addressing long-term trends, including environmental and organizational issues. The best example of a strategic-level system would be an executive support system (ESS) that employs aggregate-level data in attempts to develop long-term projections, simulations based on varying decision inputs and environmental conditions, and so forth. The goal of the ESS is to give senior managers the ability to employ a query system that lets them consider the a priori impact of strategic initiatives. For example, the head of a city planning department, in trying to determine the best route for new storm drains, could iteratively place the drain system at various locations or alter projections for precipitation and flooding in order to assess the most optimal geographic locations.

The next level down refers to management-level systems, which include information necessary for more effective monitoring and controlling of department-level activities. The two best know IS types of management-level systems are management information systems (MIS) and decision-support systems (DSS). Both systems are used for a variety of operational activities, including cost analysis, pricing and profitability analysis, production scheduling, sales region analysis, budgeting and capital investing, and so forth. DSS are often viewed as marginally higher order in terms of the processing techniques they employ through allowing managers to engage in simulation and interactive analysis. MIS tend to be used more frequently for generating routine reports and offering lower level analysis. Both types of management-level information systems enable midlevel managers to effectively operationalize or implement directives from top management through providing them with concrete and real-time information necessary to engage in rapid responses to environmental pressures or internal requirements.

At the third level in the pyramid are the so-called knowledge-level systems, including knowledge work systems (KWS) and office systems. These systems tend to be related more directly to the actual skilled work of the knowledge professionals or office staff. For example, KWS include engineering workstations, graphics workstations, and managerial workstations. A GIS professional could work at a graphics workstation, digitizing maps or conducting spatial queries, as an example of KWS activity. Likewise, the office systems most of us are familiar with include examples of word processing, statistical analysis, and document imaging.

Finally, at the lowest level of IS types are the examples of transaction processing systems (TPS) that allow for operational-level work performance. At the operational level, organizational members are concerned with summarizing transactions and specific events, generating detailed reports, sorting, merging, and updating files. They most appropriately employ TPS of the types used for work such as order processing, accounts payable, employee record keeping, and so forth. The typical TPS represents the most basic application of an organization's IS.

Managerial Decision Making

Up until now, we have made the point that IS enable managers to make better decisions by providing more complete information. However, an important point that needs to be considered is the process by which this information is normally incorporated into the decision-making process. In other words, how *do* managers use information to make decisions? What role does information play in the decision-making process that warrants such an investment in IS technology on the part of many organizations? Once we understand the part played by information in decision making, we can begin to see that IS have become an integral part of the process of effective decision making. As such, it then becomes possible to suggest (1) the *stages* in the decision-making process at which information becomes important and (2) the *types* of information that are most useful at these various points.

A phenomenal amount of research has examined the process by which managers make decisions in an effort to prescribe more efficient and effective methods. It should come as no surprise, however, that many managers make decisions in highly idiosyncratic ways. Some individuals engage in large-scale data exploration, while others make gut-feeling decisions following limited investigation (or even in seeming contradiction to the preponderance of existing information). However, when decision making is approached systematically, we can see the existence of a number of important steps. These steps often typify, in a general sense, the approach to decision making taken by most individuals. The specific steps are (1) problem recognition and diagnosis, (2) solution generation, (3) alternative evaluation and selection, (4) solution implementation, and (5) feedback (see Figure 3.3).

Problem Recognition and Diagnosis

Problem recognition and diagnosis refers to the acknowledgment that a problem exists. A "problem" simply means a difference between a planned state of events and an actual state of affairs. For example, a planned state would be a situation in which a county planning board has forecast the need for repaving 500 miles of road following winter and has budgeted money to cover that amount. However, if in the spring the board found that 750 miles of roads needed to be repaved, the county board would clearly have underestimated the costs of road repair. A problem would then exist.

In many cases, it may be relatively easy to identify (i.e., diagnose) the source of the problem and begin remedial steps. However, other problems do not lend themselves to such simple diagnosis. For example, a car that failed to start one morning could be signaling a number of different problems that would have to be checked and eliminated one at a time before the owner could be reasonably sure that the correct problem had been identified. As a result, the first step in the process of decision mak-

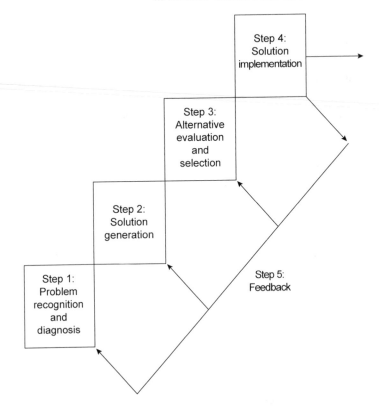

FIGURE 3.3. Steps in the decision-making process.

ing is recognizing that a problem exists and engaging in a systematic
search to make the correct diagnosis of the cause of the problem.

Solution Generation

The second step in the standard decision-making model involves the *systematic search for a series of possible solutions to the recognized problem*. For
some problems, the potential list of possible alternative solutions is seemingly endless, while in other cases the solution may be quite bounded,
either because the correct solution is obvious or because the organization
has a limited amount of resources to engage in a search for alternatives. In
the example of the county highway planning board, the set of alternatives
is rather narrow. Brainstorming may generate a solution set that would
include asking for additional funding from the county board of supervi-

sors, putting off repaving 250 miles of road until the next round of budget meetings, stretching the budget and material resources to do a superficial job of repaving the entire 750 miles, and reappropriating money from another source within the budget. It is important, however, to note that at this stage solution alternatives are only being generated, not evaluated. Prematurely evaluating the various choices often has the effect of inhibiting some individuals from taking part in the brainstorming process. The goal of this decision-making step is to develop the widest possible set of alternatives, not to make value judgments about any one suggested option.

Alternative Evaluation and Selection

After each alternative has been carefully explored in terms of strengths, weaknesses, possible ramifications, and long-term effects (if any), the next step is to *select the alternative that best meets the manager's or the organization's objectives*. The logical choice is the alternative that maximizes a decision maker's initially developed set of objectives. In other words, the best choice is often that which offers the most benefits for the least cost. However, an additional constraint is whether or not the best alternative is in conflict with other goals and objectives of an organization. As a result, assuming a complementarity between organization goals and the optimal alternative, the decision maker will then make the appropriate selection.

Solution Implementation

Once a decision has been reached concerning the best course of action, it is necessary to act on that choice. *Implementing the solution* simply means following through with a specific action or series of actions in order to solve the problem. At this point a manager must be cognizant of the potential for acceptance of or resistance to the proposed solution on the part of other individuals affected by the decision. A decision alternative may arguably be the optimal choice and still be passed over if the likelihood of its acceptance is low. Recent history highlights the fact that the Coca-Cola Company introduced its New Coke brand as a replacement for Coca-Cola Classic for what the company's management believed were correct business reasons. However, consumer backlash was so severe that the company was forced to cancel its strategic move and return to maintaining its flagship product.

Feedback

The final component of the decision-making process is *some form of feedback channel* that allows management to follow up on an implemented decision and evaluate its utility. Management assesses the degree to which the originally agreed-upon objectives are being achieved as a result of the implemented decision and, if necessary, modifies the decision or takes some other form of corrective action. Furthermore, as Figure 3.3 demonstrates, it is necessary that feedback opportunities be available at each stage of the decision-making process. If, for example, the decision makers discover that the decision is incorrect, the cause of the problem may be the result of incorrect problem diagnosis or alternative selection. Therefore, they may need to cycle back through the process and reevaluate their initial assumptions or set of alternatives in order to make a different selection.

The benefits of using an IS as a critical component of the decision-making process are obvious. First, because an IS is often a storage site for a variety of internally and externally derived pieces of data, it is invaluable in helping the decision makers to generate a selection of alternative solutions to any problem. Once a set of problems and decision parameters are specified, data can be retrieved from the system that is appropriate to the problem at hand. Furthermore, as selection criteria are developed, an IS gives managers the power to generate a set of optimal solutions that are specifically intended to maximize the alternative selection criteria. In effect, because data are stored in the IS, an organization can use its computers to process the possible solutions and then, relative to the selection criteria generated, determine the optimal solution that maximizes organization goals within the constraint of the likelihood of decision-implementation acceptance.

Finally, a computerized IS offers the best possible sort of feedback system for the decision-making process. Once solutions have been implemented and the results determined, that information forms a basis for future problem-solving activities. To illustrate: If the organization is faced with a problem that is similar to one that occurred at some point in the past, it is possible to retrieve all important information pertaining to that problem from the IS. This information would include the nature and symptoms of the original problem, alternatives generated, the solution selected and implemented, and the results of that choice. Consequently, an MIS can serve as an integral part of an organization's decision-making process, particularly when data entry is well maintained and continually upgraded.

Major Components of an IS

An IS can be used as a highly effective decision-making tool to help managers identify problems, generate alternatives, and follow up the results. Thus it is important that we gain a grounding in some of the major components of IS. Specifically, there are four basic elements or types of activities that go into the development and effective use of an IS: (1) data gathering, (2) data entry, (3) data transformation and analysis, and (4) information utilization. These categories are generally consistent with the components of a GIS, including data collection, storage, analysis, and output. Our goal is to provide the logical means to the path that converts raw data into useful information and then ultimately is incorporated in our collective knowledge base.

Data Gathering

Data gathering refers to the gathering of any and all data that may be deemed pertinent to an organization's operations. A local community, for example, would want to gather such information as census and tax-roll data regarding its residents; infrastructure data concerning roads, bridges, and other public works; planning and zoning maps for commercial and industrial development; data on optimal delivery of emergency services (e.g., fire and police); and so forth. Some data may be readily accessible (e.g., census figures), and other information may be harder to gather (e.g., possible local toxic waste storage sites created by private companies). Nevertheless, at the data-gathering stage an organization needs to determine what data are relevant to its activities and develop a plan for gathering the relevant data.

Once an organization's management team has determined which types of information can aid its operations (a potentially lengthy activity), there has to be a concerted effort to collect data. At this point a number of organizations make the mistake of assigning too few individuals to data gathering. Later discussions with managers at these same organizations reveal the managers' frustration with the seeming lack of utility of the newly developed MIS. This lack of utility is not surprising. The greater the number of organization members actively seeking and collecting information, the greater the possibility that an organization's MIS will have a significant positive effect on organization operations. It is important to remember that any computerized system is limited by the amount of data it is able to retrieve.

Data Entry

Data entry consists of any activities necessary to take the raw data that were gathered during the first step and put them into a digital form that can then be entered into computer data banks. One expensive activity that is currently taking place with a number of local and county governments is the digitizing of maps into computer-accessible forms. The old paper maps that have, for hundreds of years, been a basic tool of local governments are being converted into digital formats for computer data entry. The entry process is often long and tedious, as computer specialists convert raw data into a computer-usable format.

Data Transformation

Data transformation refers to reconfiguration into useful forms of data stored on the computer, that is, forms that are useful to managers querying the IS. For example, a number of computer software packages and structured query languages (SQL) have been developed to assist managers in recombining or restructuring data in order to provide the specific information they need. To illustrate: If a plant manager wishes to determine the relative cost level of production for one shift at his or her plant, simply accessing the raw data might be of minimal value to him or her; on the other hand, if some software packages were available to transform that data into cost structures, ratio analyses, and trend projections, the information would be much more accessible to the manager, and consequently more useful.

Information Utilization

Information utilization refers to the idea that information can be retrieved as needed by management personnel and used in making a wide variety of operational decisions. This final element in an MIS consists of the actions taken as a result of the information that has been provided to managers. In other words, information that is not relevant or useful is simply wasting a manager's time. On the other hand, the ultimate assessment of an IS utility is the effect it has on enhancing managerial decision making. If, therefore, the information that has been provided is used, it is often appropriate to judge the efficacy of an IS by user satisfaction with it (Ives, Olson, & Baroudi, 1983). In fact, to an increasing degree over the last 10 years, assessments of an MIS's impact are often measured through the surrogate of user satisfaction (DeLone & McLean, 1992; Doll & Torkzadeh, 1988; Galletta & Lederer, 1989; Igbaria & Nachman, 1990). Utilization often represents an appropriate bottom-line determinant of MIS utility. Because information is an essential element for aiding in managerial decision making, an organization's nonuse of available information offers a serious indictment of the strength of and need for an MIS.

Geographic Information and IS

In the previous sections of this chapter, we developed the role of management information in the decision-making process by addressing some of the most common types of management activity, looking at the decision-making process, and detailing the basic elements of an MIS. This section is intended to establish the basis for the relationship between an organization's overall IS and the specific role of geographic information.

Here the question that we must ask is: How is a GIS a unique component of an organization's integrated IS function—that is, what is it about GIS that makes them a unique and value-adding element in the overall IS of an organization, either public or private? As noted earlier, GIS use a computerized system for collecting, storing, analyzing, and outputting information—precisely the definition of a generalized IS. In fact, many types of IS—not only GIS—have the ability to store and retrieve information by location via use of a ZIP code, a full address, or some other spatial identifier.

What sets GIS apart from other types of IS is their ability to catalog spatially referenced objects and their attributes within the context of a map. Even more dramatic is their ability to perform quantitative analyses based on geographical principles. Thus, the GIS is different from other IS.

Within an organization, geographic information often represents a subset of the overall database. Because geographic information by nature is more specialized data, it is appropriate for and useful to a specific set of organizations, both public and private, that are involved in various activities for which geographic information is both relevant and important. In the private sector, some organizations that find geographic information useful are natural resource exploitation and development firms (e.g., mining, forestry, drilling and gas exploration companies), building and building-supply companies, farming and ranching operations, and so forth. Within the public sector, the majority of local, county, state, and federal agencies need a variety of geographic information, including resource conservation and wildlife protection, infrastructure development and repair, zoning, tax-roll updating, land management, surveying and mapping, census data analysis, and land use analysis.

It is apparent that for a number of organizations, access to and use of geographic information is not simply a luxury but a necessary and integral part of any IS that they develop. For these organizations, geographic information forms the core of their operational database. Any problems or opportunities that they seek to address must be done within the context of querying and acting upon geographic data. As a result, geographic information serves as an important element within the framework of the larger organizational IS set up to aid in management decision making (see Figure 3.4). If a GIS is analyzed in relation to the background of managerial decision making that we have explored in the chapter, it is clear that geographic information, like other forms of information in an organization's database, is useful in making optimal decisions for solving specific problems. Furthermore, the more directly an organization is tied into land and natural resource usage, the more important geographic information becomes in relation to other data sets within the overall information system.

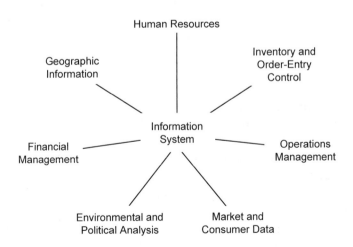

FIGURE 3.4. Geographic information as part of an overall information system.

Conclusions

This chapter has sought to provide an organizational context for the creation as well as the rapid and widespread acceptance of IS within organizations. An IS is an important and useful tool that can help managers make better (i.e., more accurate and more effective) decisions by providing them with a more complete source of information, with decision alternatives, and with possible ramifications of various choices made. Furthermore, an IS plays a key role throughout the various levels of organization management because it enables managers at all levels to access and make use of information that is specific to their operations. At higher management levels, IS are useful in organization policy and strategy development because they represent a repository of a variety of internal and external information that allows policymakers to spot trends and make necessary strategic moves. For lower level managers, an IS can provide the kind of diagnostic and prescriptive information that empowers them to operate at peak efficiency.

Geographic information is rapidly gaining in popularity and will continue to be a much-sought-after resource for a number of public and private organizations. In developing an integrated GIS, it is important that managers be aware of the natural complementarity between the geographic information that is often key to their operations and other elements in an organization's overall IS. The role of an IS is to offer managers enhanced decision-making capabilities by providing ready access to information (through the computer) and, as a result of data transformation, putting this information into usable formats that can have a timely effect on operations. In this chapter we have sought to establish the conceptual and practical link between a GIS as an organization's overall information system by placing the one within the context of the other. GIS offer particularly exciting possibilities because they represent a real expansion and sense of growth in the area of management information.

Chapter 4

Keeping the *G* in GIS
WHY GEOGRAPHY STILL MATTERS

GIS represent a departure from typical analytical and policy development tools because of their explicitly geographic component. As the use of GIS and its cognate technologies such as global positioning systems (GPS) and remote sensing becomes more common, the potential of these tools will become more readily apparent. One of the keys to unlocking the potential of GIS is a better understanding of geography and spatial analysis among GIS users. This chapter attempts to shed light on the importance of the principles of geography, cartography, and spatial statistics to GIS implementation using examples drawn from public policy analysis and cartography. It recommends that organizations implementing a GIS develop a three-point approach to maximize the chances for success: (1) substantive expertise in the field of application, (2) knowledge of GIS techniques, and (3) understanding basic principles of geography, cartography, and spatial statistics.

GIS represent a significant departure from typical analytical and policy development tools because of their explicitly geographic component. As the use of GIS has become more widespread, the potential of this tool in its various applications has become increasingly apparent. However, in its current state of development, given the level of geographical knowledge and understanding of most users, there is still room for improvement. The key to unlocking the full potential of GIS lies in a better understanding of geographic, spatial analytic, and cartographic principles among GIS users.

Taking examples from the geographic literature, in this chapter we identify several problems that average GIS users may encounter while employing the technology and describe these problems as a means to highlight the importance of geography, cartography, and spatial statistics to the successful implementation of GIS. This chapter starts by covering two topics: the importance of scale and aggregation in spatial analysis, and the appropriate development and use of maps, especially with regard to generalization. It continues with a discussion of the advantages of using spatial statistics within the context of GIS to gain even more insights using this tool. It concludes with a three-point approach to implementing GIS for analytical purposes, suggesting that the presence of substantive expertise, knowledge of GIS techniques, and the understanding of basic geographic and cartographic principles will improve the chances for an organization's acceptance and use of a GIS.

Background

Although GIS have a history that dates back to the 1960s, we are still a long way from unlocking the full potential of GIS as an analytical tool, especially when it comes to the use of spatial statistics. In the early 1990s, Goodchild and Getis (1991: 1–10) acknowledged that while the potential for GIS is vast, the orientation of GIS applications at that time leaned toward information and infrastructure management rather than toward spatial and policy analysis. In essence, as the authors noted, GIS is to spatial analysis as the statistical software packages (e.g., Statistical Package for the Social Sciences [SPSS]) are to traditional statistical analysis, and thus represent a powerful tool indeed. More than 10 years later, Anselin (2002) reiterated this point, indicating that average GIS users are still a long way from using the full analytical capabilities of GIS. On the other hand, Sawicki and Peterman (2002) suggest that there is a small but growing contingent of users, many of them nongovernmental organizations, that are using GIS for analytical purposes.

Even today, most real-world applications of GIS rely primarily on the information storage, retrieval, and management functions of the technology. While there is nothing inherently wrong in using GIS in this way, the technology can be used to perform far more complex tasks in addition to the basic tasks of storage and retrieval. The key to using GIS to perform more complex tasks lies in an understanding of the basic geographic and cartographic principles that underlie spatial analysis and modeling, as well as data input and analytical output.

Geography is important to GIS for two major reasons. The first reason is that a sound understanding of basic geographic principles will increase the likelihood that GIS users will employ the technology in a manner that is appropriate, logical, and ethical. The second reason is that the current emphasis on information storage and management undersells the value of GIS as an analytical tool. By way of analogy, the continuing emphasis among GIS users on the more basic aspects of GIS has the effect of using a Cray supercomputer to solve simple math problems, processes that are well within its capabilities but not the least bit challenging; or owning a Maserati yet using it only to drive to the corner grocery store. A more prominent role for geography will begin to unlock additional analytical capabilities, and thus make the technology even more powerful.

We should not be surprised that the geographic potential of GIS is relatively underutilized. Many reports appearing in both scholarly journals and the mass media over the past decade have described the U.S. populace as geographically illiterate, and have provided examples of the inability of students at all levels to identify and locate on a map what should be familiar places. Indeed, singling out map identification as an indicator of geographic literacy itself overlooks the breadth and depth of geography as a discipline and ignores geographic principles as a foundation for the development of the field. However, if Americans as a nation have difficulty with basic map identification, how can we expect them to be knowledgeable about principles of geographic location?

Basically, geographic principles (and thus geography) help to explain and predict the location of people, places and things. Consistent with this description of geography, Goodchild and Getis (1991: 1–2) define spatial analysis as "a set of techniques whose results are dependent on the locations of the objects being analyzed." That is to say, spatial analysis is appropriate in situations wherein when the objects move, the results change. The authors give as an example the U.S. Center of Population, which has been moving westward over the years, following the migration of people within the country from the northeast toward the west and southwest over the past two centuries. When the Founders established the United States, its center was located near the Eastern Seaboard; today, the mean U.S. center of population is in south-central Missouri (U.S. Census Bureau, 2007).

Another example of spatial analysis can be found in the impact of the movement of people in the United States on the location of congressional districts and the allocation of members of Congress among the states. For example, after the 2000 decennial census, the Sunbelt states of California and Texas added districts and representatives, while several Midwestern states, such as Iowa, lost seats.

It is important to note, however, that many of the statistical analyses with which most of us are familiar are unaffected by movement: they are aspatial. For example, the average income of the U.S. population is unaffected by the states of residence of U.S. citizens (Goodchild & Getis, 1991: 1–2). This statistic is based on aggregate figures.

Whereas the current orientation of GIS in practice is on information storage and retrieval, especially emphasizing information related to infrastructure and natural resources, GIS are capable of far more complex tasks. Among these tasks are map measurement, particularly area measurements, which are simple to conceptualize but difficult to execute manually (Goodchild & Getis, 1991: 1–12). Such tasks are among common capabilities of contemporary GIS software. But GIS can also be used to perform even more challenging analyses because of their vast capabilities for data integration and manipulation. Such analyses include those needed in the development of global science (Goodchild & Getis, 1991).

The point to bear in mind, however, is that in order to maximize the use and productivity of GIS, it is necessary to understand basic geographic and cartographic principles. The following sections suggest how a sound understanding of these principles can enhance the use of GIS for analysis, and why this is valuable.

Analytical Examples

Tobler's First Law of Geography boldly asserts that "everything is related to everything else, but near things are more related than distant things" (Tobler, 1970: 236). This is the basic assumption that underpins geography as a discipline. Geographers devote their professional lives to looking for relationships and causal links between and among people, places, things, and events on Earth. These relationships and links take the form of geographic or spatial patterns that cover the planet on, under, and above its surface. For example, when we read newspaper accounts about the migration of people or jobs to the "Sunbelt" or flooding in the "Midwest," we have a common, but general, understanding of what region the reporter is describing. Our understanding is based on our recognition of relative uniformity within the region and some contrast between the regions.

Specifically, the word *Sunbelt* conjures up images of—What else?— warm, sunny climes, leisure activities, retirees, and, in the last decades of the 20th century, job growth. The term *Midwest* brings to mind amber (and green) waves of grain and other crops, solid American values, the pioneer spirit, and, in recent decades, job loss. However, in spite of the similarity of our generalized images of these regions, when we get down to specifics, differences of opinion arise and may become heated.

Take the case of circumscribing the Midwest. There is probably widespread agreement about including the states of Illinois, Indiana, Iowa, Michigan, Ohio, Wisconsin, Missouri, and Minnesota. Some people argue, however, that Kansas, Nebraska, and Oklahoma should also be included. Others suggest that South Dakota, Kentucky and/or Pennsylvania (sometimes in whole, sometimes in part) should be included as part of the Midwest. I have never forgotten a long-ago conversation with a sociologist friend during which he defined the Midwest as stretching "from Cleveland to Colorado." In the geographical literature, Zelinsky's (1980) Midwest includes all or parts of the states of Ohio, Michigan, Indiana, Illinois, Wisconsin, Missouri, Iowa, Missouri, Minnesota, South Dakota, Nebraska, Kansas, and Oklahoma. By contrast, DeBlij and Muller (1992) consider large portions of the states of Illinois, Indiana, Ohio, and Michigan to be more appropriately classified as belonging to what they call "The Anglo-American Core," which by their definition is "synonymous with the American Manufacturing Belt" (215). Finally, Garreau (1981) includes most of Illinois with "The Breadbasket," but classifies much of Indiana and most of Ohio and Michigan as parts of "The Foundry." Southern Indiana and Illinois, by the way, are part of Garreau's "Dixie." And then there is a more recent variant of the term: "Heartland." (Or perhaps it is only wishful thinking on my part, as a native Midwesterner, that these terms are more or less synonymous.)

How can it be that a region that we talk about so casually can be so difficult to define? There are two major reasons, the first relating to the difficulty in developing meaningful taxonomies or classifications, the second relating to Tobler's First Law of Geography.

What we are seeing in the classifications of Zelinsky, DeBlij and Muller, and Garreau is their idiosyncratic definitions of what constitutes a region. Gould (1983: 72) observes:

Taxonomy is often regarded as the dullest of subjects, fit only for mindless ordering and sometimes denigrated within sciences as mere "stamp collecting." . . . If systems of classification were neutral hat racks for hanging the facts of the world, this disdain might be justified. But classifications both reflect and direct our thinking. The way we order represents the way we think.

It is likely that the three classifications of "midwestern" states described above reflect three different sets of criteria for classification that the three different authors have chosen to define their regions. We are seeing differences in the way the authors think, reflected in the ways that they order the criteria defining regions. Anyone who has driven throughout the Midwest region can testify to the existence of miles and miles of productive farmland interrupted by factories, some of them still in operation, others long closed. Human settlements in big towns and small, along with natural open space, complete the view. Given individual differences in the perceptions of the individuals doing the classifying, we should not expect an exact match in the development of a regional classification scheme. The way they order represents the way they think.

However, whereas Gould's explanation gives us a general clue about why there can be so much difference in how three classification schemes could be so different, Tobler's First Law of Geography provides much more specific insight. To reiterate, Tobler asserts that "everything is related to everything else, but near things are more related than distant things." Tobler's law gets to the heart of an issue of crucial importance to geography: scale.

Several examples will help to illustrate Tobler's law, which applies to both history and geography. Taking a historical example, it has been noted that a weather forecaster would have a very good probability of predicting tomorrow's weather if he or she did nothing more than say that today's weather pattern will continue through tomorrow. However, if that same forecaster predicts that today's weather pattern will prevail at a date 6 months from now, the likelihood of an accurate prediction would be minuscule, particularly in a region like the midwestern United States where temperatures can fall well below zero in the winter and soar to the 100-degree (Fahrenheit) range in the summer. (A favorite saying in the Midwest is "If you don't like the weather, just wait a few minutes.") And while predicting today's weather pattern for a date 3 months, 1 month, or even 1 week from today might be closer to what we would actually see on those dates than the weather we might see on a date 6 months away, the probability of accuracy would not be nearly as great as the initial prediction that tomorrow's weather will be like today's.

Similarly, in the geographical mode, we can predict with a high probability of accuracy that today's weather in the town nextdoor will be the same as the weather in our own hometown. As we begin to predict our weather pattern as the prevailing pattern for another state, the likelihood of accuracy declines. For example, in Indiana, we might well expect that during a drive to Illinois or Ohio we would experience little or no change in weather. On the other hand, as our travels continue, and we arrive in Maine, we might very likely see a significant difference in temperature, precipitation, and humidity. At the extreme, the difference between a typical hot and humid Indiana summer's day and the frigid Antarctic on the same day show the folly in trying to extrapolate over large geographical areas.

We need not confine our examples to the physical. For example, consider the cultural differences we see among people of various places. While Hoosiers (natives of Indiana) may seem a lot like Buckeyes (natives of Ohio), they will appear to be less like Mainers (especially members of one of the four Native American tribes in that state), even less like the French, the Japanese, or the Maoris of New Zealand.

And yet, consistent with Tobler's law, close to Indiana, we would see many similarities. And even at great distances, we could find relationships between places. For example, the hole in the ozone layer over the Antarctic is usually attributed to the activities of people in industrialized countries. Hoosiers and Buckeyes, distant as they are from Antarctica, affect it.

Embedded within Tobler's insight that everything is related, but nearer things are more closely related than distant things, is the hint that transitions occur across space. These transitions are rarely smooth and clearly defined, however. A visit to Maine demonstrates this fact. Maine is located adjacent to the Canadian province of Quebec, where French is the official language. Most Quebeçois (residents of Quebec) are descendents of French immigrants to the region. In spite of the fact that there is very clear and definite line on the map representing the political boundary separating Maine from Quebec, the cultural "line" or transition on the ground is, as Goodchild (1988) might describe it, "fuzzy." A large proportion of the Mainers who live near the border of Quebec are self-described "Francos," that is, Mainers whose ancestors are French. Many of them continue to speak French as their first language. And although the concentration of Francos declines as one leaves the Quebec border and drives "downeast" to the Maine coast, the descendants of French immigrants can be found throughout Maine and recognizably into Massachusetts. By the time one gets to Indiana, people of French ancestry are few and far between, having been replaced by descendants of British, German, Eastern European, and African peoples.

The point is that drawing a political boundary is a relatively easy task once the border countries agree on who gets what. Such a border will have exact geographical coordinates. On the other hand, trying to provide an accurate, understandable, meaningful map showing the attributes of the land and its people is much more difficult because of the problems created by transitions, such as the one described above. So-called fuzzy borders are the result (Goodchild, 1988). So far, there is no easy way to represent fuzzy borders on a map. Perhaps one day improved analytical capabilities in GIS will help resolve this problem.

It is not only cultural transitions that produce fuzzy boundaries. The natural environment abounds with similar transitions. Patterns in vegetation, soils, wildlife, and many more things are examples of transitions in the natural world.

The problems of representing "ground truth" (what objectively exists in reality) are well known to cartographers. Imhof (1963: 14) describes cartography as "a technical science with a strong artistic trend." He continues:

> But theoretical cartography is not yet sufficiently characterized. The motif or theme of the map is the earth's surface. The essential element of any drawing (or pictorial design) is observation. The people who are drawing the surface of the earth must observe and study it. The geographer also does this; this is a geographical task. In this respect, there is a very close relationship between geography and cartography. To a certain degree the cartographer is a geographer; he is a graphic geographer or a geographic artist. As an applied science, theoretical cartography bridges the connection between techniques to art, to all the different sciences that are concerned with the study of the phenomena of the earth's surface.

Clearly, the cartographer's task is not an easy one. And in "the olden days," when a cartographer's tools emphasized pen and ink, most people found the cartographer's craft daunting. Few took the challenge to learn the intricacies of cartography. Nowadays, however, the introduction of GIS and computer-assisted design (CAD) and computer-assisted mapping (CAM) technologies systems enables nearly anyone to make a map.

It would be a big mistake, however, to assume that the ability to make a map makes one a cartographer. Even as long ago as the 1960s, Imhof raised concerns about the use—and potential misuse—of "technical aids" in the hands of people who are unfamiliar with the basics of cartography and geography. He noted that "the cartographic problems concerned with the graphic composition and design of the map have increased in complexity, in spite of the new technical aids" (1963: 15). Imhof identifies five problem areas in modern cartography: generalization, coordination, the visual effect, abstract and direct pictorial representation of terrain, and thematic mapping.

Generalization

All maps are generalizations of the areas they represent. Imhof notes that even if we were to take an aerial photograph of an area, then reproduce it at the size we desire, we would find it difficult to identify all the relevant features for a specific task. It is the cartographer's task to "transform, emphasize, eliminate, summarize, exaggerate and enlarge certain things" in the map to convey an image that is usable and tells a story (Imhof, 1963: 17). This is the process of generalization. As Imhof (1963: 17) goes on to say, "The solution of a mathematical task is either right or wrong, but the solution of a cartographic task can be evaluated only in degrees of good or bad." At one time or another, most of us have had experiences with maps that were inaccurate, incomplete, or ambiguous. Perhaps the best example of such a map is one that a friend draws by hand to help you find your way to his or her home. If you are using such a map to get from one place to another, you may find key landmarks missing, roads drawn where none exist in real life, or other points of confusion that make using such a map frustrating. Perhaps you have drawn such a map yourself. Generalization must be done with great care, keeping in mind both the purpose of the map and the audience for which it is intended.

Coordination

Once the cartographer has chosen the features he or she wishes to represent and a way to represent them, it is necessary to coordinate the resulting elements. The idea is to create a map that conveys information in a form that is usable and informative for its intended audience. This means that roads, rivers, and political boundaries must be distinguishable one from the other. It means that topographic features, where present, must be recognizable. It means that the names of towns, cities, states, and other places must be placed so as not to obscure other important information, and yet be placed in a way so as to be unambiguously related to the correct place. As Imhof (1963: 17) notes, it is possible to design a map with excellent elements only to find that the neglect of the overall effect results in a poor map.

The Visual Effect

Closely related to the problem of coordination is the problem of visual effect. What does the map look like once it is completed? Is it neat or cluttered? Are the symbols helpful or confusing? Does the map convey the message it was designed to convey? In assessing the visual effect of a map, it is crucial to keep in mind that a map is designed to convey information or ideas to a specific audience. It is not enough that the mapmaker understands his or her own creation. The visual effect must be clear to the audience for which the map is intended.

Different maps have different audiences. A good example of different maps for different audiences may be found in the *Rand McNally Road Atlas of the U.S. and Canada*. The basic *Rand McNally Road Atlas* includes a detailed map for each of the states and provinces of the United States and Canada. In planning a road trip for my family, I discovered a similar map, also published by Rand McNally, that provides the same map coverage, but at a greater level of generalization. I purchased this version for my elementary-school-age children to use as we made our trip. Later, I noticed that the cover features a clearly prosperous and active, but undoubtedly grandparently, couple, leaving no question about the audience for this "large type" edition of the atlas.

Abstract and Direct Pictorial Representation of Terrain

The representation of terrain is an important aspect of every map. Imhof notes that such representation "forms the look of many maps more than anything else" (1963: 18). The representation of terrain on maps must fulfill two requirements: first, it must represent surface shapes geometrically, and second, it must present a readily recognizable pictorial representation of the terrain (Imhof, 1963). Typically, contour lines are used to show variations in elevation. Unfortunately, elevation lines require careful inspection to reveal the lay of the land. Specialized training and practice in reading elevation maps also comes in handy. Most people would find interpreting these maps a challenge. In order to overcome this difficulty, cartographers often use shading and other more obvious visual devices to render a picture that more obviously depicts hills and valleys for the more casual map reader.

Whether the cartographer uses contour lines or shading, Imhof notes that "any cartographic terrain representation are [*sic*] somewhat fictitious in nature" (1963: 14). In part, the fiction is related to the fuzzy boundary and generalization problems that make it necessary to draw a fixed line to separate one elevation from another, when the transition is far more likely to be gradual. The other part of the fiction is the image, created by shading, hachures, or use of colors, that the map reader sees and visually understands. This image may or may not accurately represent the terrain. What you see may or may not be what you get.

Problems of Thematic Mapping

Finally, Imhof describes problems of thematic mapping. Thematic maps are typically designed to show the existence of a specific object or attribute across a defined and limited area on the surface of the Earth. Imhof gives the example of population. In some maps of this type, the cartographer places circles, dots, or some similar symbol of various sizes to correspond with the geographic location of cities, towns, or other places. (This kind of map uses "graduated symbols.") When a map of this type is well done, the map reader gets a good idea of the sizes of various places.

Another method frequently used in thematic maps is to show the incidence of particular objects or attributes through the use of shading or a color ramp. The usual rule in such a practice is the higher the density of the object or attribute in a unit of land, the darker the shading in that area. A serious problem may arise in using this method. Because the areas into which the map is divided for shading purposes are typically administrative or political districts (states, counties, etc.), the resulting image may be misleading. For example, if one were to take the cities of Illinois and shade them according to their level of poverty, the city of Chicago (and most cities) would be more or less in a middle range. Yet a drive through Chicago quickly reveals that there are vast differences in the wealth of city residents, from the millionaires on the Gold Coast of Lake Michigan to the poverty of the Cabrini Green housing project nearby. The typical shading found on many thematic maps causes the loss of much important detail. Imhof contends that many such maps exhibit cases of "shocking ambiguity" (1963: 21).

The dilemma with the use of a GIS is the fact that it allows any reasonably intelligent, computer-literate person to make a map, regardless of his or her knowledge or lack of knowledge about the problems associated with cartography. Even more seriously, an individual without cartographic training who sits down at a GIS workstation to make a map may very well be unaware of his or her limitations as a mapmaker. The ease with which a person armed with a GIS can make a map may lull the individual into a false sense of security in expertise. It is important to realize, however, that limitations of expertise do exist. Just as no reasonably intelligent person who has mastered Word on his or her personal computer would automatically assume that he or she has suddenly become capable of writing a novel with the staying power of *A Tale of Two Cities,* neither should the novice who has mastered the use of a GIS assume that he or she has suddenly become a cartographer.

In trying to describe the basic problems inherent in undertaking geographical analysis and cartographic tasks, we have provided some general examples of the difficulties that one might encounter. Below, we provide examples designed to illustrate the complexity of geographical analysis and cartographic design and use. The examples we present come from the public arena, but appropriate examples can be found in the private sector as well. In the public sector, the potentially widespread public impacts of analysis and resulting policy choices pointedly reinforce the importance of a sound understanding of geographic, spatial analytic, and cartographic principles within the framework of a GIS implementation.

Regional Equity in Chicagoland

The first example emphasizes the complexity of spatial analysis, especially as it relates to scale, and comes from Chicago and its surrounding counties, which are united by a jointly funded transportation system, the Regional Transportation Authority (RTA). The example recounts a debate among the members of the RTA about "regional equity," a policy theme centering on the geographical sources of revenue versus the geographical distribution of funding for transportation services within the RTA's service delivery area (Obermeyer, 1990b).

The RTA was formed in 1974 by the Illinois legislature and won approval in a state referendum. The purpose of the RTA was to expand the tax base beyond the City of Chicago in order to support public transportation in the region, which covers Chicago and Cook County, along with the "Collar Counties" of DuPage, Kane, Lake, McHenry, and Will.

Funding for the RTA comes from three sources. The first of these is farebox receipts. The second is federal funds, such as those provided by transportation grant programs. The final source of funds is sales taxes collected from throughout the region. A crucial point in analyzing regional equity in this example is that the sales tax rate is not uniform throughout the region: Cook County residents pay a sales tax of 1% to support the RTA, while Collar County dwellers pay only one-quarter of 1% (0.25) in sales tax.

Originally, the RTA directly operated three separate transportation services under the auspices of three different service providers: the Chicago Transit Authority (city buses and rapid transit), a commuter rail network, and a suburban bus system. A budget crisis in 1981 resulted in a reorganization of the RTA that formally separated the services and made the RTA an administrative, rather than an operational, body. After the reorganization, the Chicago Transit Authority continued to be known as the CTA, the commuter rail network became "Metra," and the suburban bus system became known as "Pace." The RTA became an administrative overseer, rather than a direct service provider.

There were two major sources of geographical conflict in this case. The first source of conflict was the uneven density of transportation services in the metropolitan service delivery area. Transportation service is much more readily available in the City of Chicago and Cook County than it is in the Collar Counties. The second conflict arose because Collar County residents perceived themselves as "forced riders," paying for a service for which they had relatively little use, as evidenced by low ridership among suburban residents of these outlying counties.

The regional equity debate began within the context of the development of the RTA's budget for fiscal year 1986 as the transportation services jockeyed for position in order to maximize their shares of the sales tax revenues collected by the RTA. Metra initially raised the issue, arguing that the suburbs provided 60% of the RTA's funds from sales taxes, while receiving less than 40% of this revenue to cover the operating expenses of the two services providing public mass transportation in the suburbs: Metra and Pace. Of course, Metra's suggested solution to this problem was greater funding for themselves and Pace.

The CTA justified its relatively larger share of the sales tax revenues by noting that 37% of the trips taken by suburbanites were via CTA vehicles, while only 21% of the sales taxes collected in the suburbs to help fund the RTA went to CTA operations. Pace responded by formally declining to become involved, and suggested that the regional equity theme was a smoke screen, that equity is in the eye of the beholder, and that Metra's primary objective in raising the issue was to maximize its share of RTA revenues.

Both Metra's and the CTA's analyses were based on verifiable facts and were therefore valid in this respect. The difference, however, is that each relied on a different set of review criteria focused at different scales. Metra's analysis was valid at the regional, aggregate scale. The CTA's analysis was essentially focused at the local level. Not surprisingly, each arrived at a different solution.

We can see the difference that scale of analysis makes even more clearly by comparing Metra's regional aggregate analysis with an analysis based at the individual disaggregated level. To reiterate, Metra's analysis charged regional inequity based on the fact that the suburban areas of metropolitan Chicago contributed 60% of the RTA's operating funds through sales taxes collected, while the two suburban-based transportation services received less than 40% of the RTA's revenue from sales taxes to support operations. At the individual level, we have a starkly different situation: suburban residents pay only one-quarter of one percent (0.25%) in sales taxes to the RTA, while City of Chicago residents must pay one full percent (1%) in sales taxes, four times the amount required of suburbanites. In this instance, changing the scale of the analysis from the regional aggregate level to the individual disaggregate level changed the results and conclusions of the analysis dramatically.

Of course, it is important to point out that the idea behind the creation of the RTA was to expand the public transportation tax base to include suburban areas. Based on its operating charter, it may be argued that it was never intended that each geographical area should get back in services exactly what it had paid in taxes: if that had been the intention, there would have been no need for an RTA. At the very least, the funding and spending mechanisms would have been designed differently (Obermeyer, 1990b).

If we carry this theme to its logical conclusion, we must also consider the effects of scale on the development and implementation of policy. An excellent example of this theme is found in Paul Knox's (1988) article "Disappearing Targets?: Poverty Areas in Central Cities." In his examination of poverty areas within U.S. cities, Knox documented the existence of variations in both the nature and the intensity of urban poverty, attributing these differences to the "differential imprint of economic and demographic change among cities and regions of the country" (502). He suggests that the system established under the Reagan administration and maintained throughout the first Bush administration treated poverty as an individual problem, and delegated to the states the responsibility for providing Aid for Families with Dependent Children (AFDC) payments to the poor living within their borders. As a remedy, Knox recommended that public policy be targeted spatially and be "customized to address the problems inherent to specific, localized cycles of deprivation" (506). Such targeting might include locating well baby and family medical clinics and job training and development centers in the neighborhood or providing special incentives to businesses to locate in these areas as means to enhance local economic opportunities.

The more general theme of Knox's article is that how we define problems influences how we address them (an idea consistent with Gould's (1983, 1994) point that the way we order reflects the way we think) . More specifically, Knox suggests that the scale at which we define problems determines the strategies we adopt to resolve them. For example, the Bush administration rejected the Kyoto Protocol as a drain on economic development. Contrarily, in 2005, the Republican governors of California and New York and a growing number of firms began to support cutting greenhouse gas emissions. One justification for this newly "green" position is

> to save money, particularly since cutting greenhouse-gas emissions usually means increasing energy efficiency, according to [J. Edward] Hoerner of Redefining Progress. The organization released a report last month showing that California residents and businesses would save enough money to create some 65,000 new jobs through the spending boom that would result from cutting carbon dioxide emissions. (Kelly, 2005)

By February 2007, following on the heels of George Bush's State of the Union Address in which he acknowledged the problem of global warming, a group of multinational firms operating in the United States, including General Electric, endorsed a formal statement to fight for clean energy. Among the reasons given are the need for predictability in policy, and because it is a "moral issue" (Milton, 2007). More cynical observers note that General Electric had quietly become a world leader in windmill technology, and thus stood to benefit economically from the shift.

While GIS give us enhanced operating capacity in terms of the increased quantity of data that we can analyze, they cannot by themselves develop appropriate models and scales of analysis in a specific circumstance. That is for the project analyst or manager to do. It is therefore necessary that individuals responsible for developing approaches to analyzing geographic information have a sound understanding of geographical principles in order to prepare them to perform this assignment competently.

Maps and Their Appropriate Use

A second area of concern that we examine is the appropriate use of maps and the concomitant need for cartographic expertise. This theme is worthy of attention because of the important role that maps play both as inputs into and outputs of GIS. As inputs, existing maps may be scanned or digitized to provide a basic geographic framework for other data in a system. As outputs, maps may be used by a variety of people for a number of different purposes, some of which may not have been intended by the creators of the map. More seriously, some of these unintended uses may be entirely inappropriate.

In his book *How to Lie with Maps* Mark Monmonier (1991, 1998) provides a thorough discussion of the use and misuse of maps. Specifically, Monmonier notes that because of the availability of personal computers and other electronic map-making technologies, map making has become available as a tool to anyone who has the price of the hardware and software, or who finds him- or herself assigned to this task on the job. Not surprisingly, many of the people who now assume the role of cartographer, or find themselves thrust into that role, have no training whatsoever in cartography. Consequently, the probability of inappropriate map creation or use has increased in recent years. Certainly, the proliferation of GIS exacerbates this situation by increasing the ease and speed of map creation and use.

This is clearly a problem. Monmonier (1991: 2) emphasizes the seriousness of the problem:

> *[A] single map is but one of an indefinitely large number of maps that might be produced for the same situation or from the same data.* The italics reflect an academic lifetime of browbeating undergraduates with this obvious but readily ignored warning. How easy it is to forget, and how revealing to recall, that map authors can experiment freely with features, measurements, area of coverage, and symbols and can pick the map that best presents their case or supports their unconscious bias. Map users must be aware that cartographic license is enormously broad.

When using GIS to make new maps as well as when using existing maps to create other maps, caution is in order. Both inadvertent mistakes and intentional bias may make a specific map unusable for a variety of purposes—perhaps even for the purpose for which it was intended. However, even mistake-free and unbiased maps may be used inappropriately.

An excellent example of this situation is discussed by Gersmehl (1985) in an article in which he describes how one of his own maps was misused. In 1977, Gersmehl published a set of coarse-resolution dot maps showing the general distribution of soil orders in the 48 contiguous states of the United States. His map showed the presence of a general class of soils, histosols, in several western states. These histosols represented some widely dispersed peatlands as well as "muck."

Although they both qualify as histosols, peat and muck are different: peat is valuable as an energy resource, muck is not. Herein lay the basis for misinterpretation and misuse of Gersmehl's map.

Attempting to identify potential energy resources in the United States, the U.S. Department of Energy (DOE) came across Gersmehl's map and included the information provided on it in the agency's map of commercial energy fuel. In compiling this map, the DOE failed to recognize that the class "histosol" was a very broad soil category that included everything from peat to muck. Instead, the agency classed all the histosols as peat, an assumption that resulted in a map showing a much greater amount of this valuable energy resource than actually exists. The result was a map that was factually incorrect through no deliberate intention to falsify, either on the part of the creator of the original map or the developer of the derivative map.

Gersmehl (1985) related his surprise and concern about the misuse of his map and concluded:

> A person who puts information on a map has a duty to be fair to the data, to be clear to the map reader, and to try to anticipate the ways in which a third person may be affected by a foreseeable interpretation of the map. (333)

This caveat seems to put all the responsibility on the cartographer not only to make a map that is accurate and readily understandable, but also to try to anticipate the many ways in which a specific map might be misused. Unfortunately, Gersmehl's emphasis on the cartographer seems to let the map user off the hook. Not only must the map*maker* be aware of the potential misuse of maps, but the map *user* also bears responsibility for using existing maps appropriately. This requires some basic understanding of cartography along with a willingness to make the effort to consult with the original cartographer regarding appropriate use of a specific map if there is even the slightest question about the information transmitted via the map.

If caution is appropriate in the use of paper maps, then caution is most certainly required in the use of GIS. The ease of use of the technology, coupled with its speed, means that GIS can encourage the proliferation of cartographic errors at a rapid pace in the hands of inexperienced or unknowledgeable users.

One development designed to address this problem has become a fixture of GIS since the first edition of this book: metadata. Metadata is data about the data that is provided as part of the GIS. Metadata include the following information about data: date of collection, who collected it, the scale and projection of any maps, and so on. From this information, a user should be able to make some assessments of the usefulness or validity of the data for his or her specific needs.

It is important to note that metadata do not magically appear. In developing a GIS, the people responsible for the project must insert this information. Most state-of-the art GIS software provides a convenient framework for including the metadata as part of the GIS. Still, the hard work of gathering the information remains, and must be done. As well, users of the GIS must review metadata to make sure that the data are appropriate and current for their purposes. Chapter 10 covers the topic of metadata.

A Three-Point Approach to Responsible GIS Application

The importance of GIS as a tool in policy analysis is growing. With improvements in the speed and capabilities and reduction in the price of this technology, its importance is likely to continue to grow. The preceding discussion of how changing the scale and aggregation of an analysis affects the results highlights the critical importance of an understanding of geographical principles and at least rudimentary spatial analysis before undertaking such complicated tasks. Similarly, the potential for misuse of maps suggests that a GIS user must understand the basic principles of cartography. In short, GIS may be used or misused. An appropriate balance of skills and knowledge is the only hedge against misuse and is the key to unlocking the full potential of the technology.

In order to unlock the full potential of GIS, as well as to use the technology in an appropriate manner, we recommend a three-point approach to applying GIS effectively. In order to maximize the functions of the technology, GIS users must develop the following three attributes: (1) substantive expertise in the application in which the GIS is to be used, (2) knowledge of GIS techniques, and (3) an understanding of geographic and cartographic principles, at least in rudimentary form.

Each application of GIS is unique. To begin with, each application occurs within the framework of a specific organizational mission and has specific geographic boundaries. Therefore, for example, using a GIS to identify potential toxic waste dump sites throughout the United States is very different from using a GIS in the City of Cincinnati for infrastructure management. Each application requires specific substantive expertise related to the organizational mission and the specific task at hand.

Presumably, the substantive expertise already exists within the organization adopting a GIS. It is critical to make use of that expertise, to find a means to keep the organization's substantive experts in the GIS loop first to help develop data specifications and sources, then to monitor the implementation, and finally to evaluate the results. These substantive experts are in the best position to determine the validity of results and recommend necessary changes.

Knowledge of GIS techniques and specific knowledge of how to operate the system chosen is, of course, absolutely necessary. The individual in the GIS driver's seat must know not only the commands to operate the system, but must also know both the capabilities and the limitations of the system and the data on which the system relies. It is his or her job to assure that all tasks undertaken using the GIS are within the capabilities of the system.

Finally, the importance of understanding basic geographic and cartographic principles cannot be overemphasized. Critics have argued that this third point is exclusionary. That is, it sets up a standard that allows only geographers to handle an organization's GIS. It is not our purpose to recommend a kind of technical "litmus test" for using GIS technology. We argue instead that whomever an organization charges with the responsibility for operating its GIS must have sufficient geographic and cartographic background so that problems of the type mentioned in this chapter will not interfere with the successful application of GIS technologies.

But how can an organization making use of a GIS assure that it meets all three of these requirements? Modern education, with its emphasis on specialization, virtually assures that no single individual will possess substantive expertise in the organization's function, knowledge of GIS techniques, and an understanding of geographic and cartographic principles.

More likely is a situation wherein these three areas may be covered by two, three, or perhaps even more different people.

Obviously, when several individuals collaborate, this requires extensive and careful coordination to make certain that all relevant professionals are kept in the loop. Such coordination begins with the realization and acknowledgment that all three functional areas are of equal and critical importance. Without the active involvement of each area and mutual respect between and among the experts, there is a danger of an unsuccessful implementation. Dominance by each of the areas brings its own particular difficulties.

For example, if the substantive experts fail to acknowledge the expertise of the GIS professional, they may insist on operations that are inappropriate for the system or database available to them. If they do not respect the knowledge of those with expertise in geography and cartography, they may make mistakes related to scale or cartographic features and encourage outcomes such as those described in Gersmehl's study.

Similarly, if the GIS expert fails to respect the substantive expertise and geographic and cartographic knowledge of his or her colleagues, he or she may fail to listen carefully to the information and recommendations they offer and consequently fail to implement them properly in the system. This can result in a system that is poorly or incorrectly specified, and consequently will not perform up to its potential. Anyone who used computers in the days of punch cards and the all-knowing "consultants" has experienced this potential problem firsthand.

Finally, if the person who is knowledgeable about geography and cartography does not respect the expertise of the substantive specialist and the GIS professional, he or she may miss critical details that are unique to the specific application involved. Again, the project will suffer.

Each organization will have to develop its own strategy for coordinating these three groups. Regular meetings and discussions may help promote communication and cooperation. In addition, providing training and education via workshops or enrollment in relevant courses in nearby colleges and universities or online may also help the individuals involved gain a basic understanding of all three elements of a project.

Conclusions

Effectively applying a GIS technology is not an easy task. It requires an understanding of the substantive application area, a thorough knowledge of the GIS software employed, and an understanding of the basic principles of geography and cartography. Of these three components, the third is least likely to be recognized as necessary. And yet, as we have seen, it is crucial to the development and successful implementation of a GIS used for analytical purposes.

When organizations think about implementing a GIS, their first thought is to acquire a system and hire a technician or train an existing employee. Under these circumstances, it is no wonder that spatial analysis is underused in GIS, as Anselin (2002) reminds us. On the other hand, failure to use GIS for spatial analysis means that the GIS may not be used as fully as possible, and therefore the organization may not accrue the full benefits of owning the technology.

In order to achieve the full benefits of owning a GIS, it is necessary to have substantive expertise, GIS skills, and understanding of basic geographic and cartographic principles. While we can only suggest how this triumvirate of knowledge may be achieved, each organization must find its own way. The first step, however, is recognizing the need.

Chapter 5

GIS and the Strategic Planning Process

The purpose of this chapter is to demonstrate the integration of geographic information into an organization's overall strategic planning process. We show that a GIS can be used as an important and central element in the development of strategic objectives and long-range plans. We define the concept of strategic planning and we propose a general model of strategic planning that will serve as the basis for gaining a better understanding of all relevant elements in the strategic planning process. Finally, we analyze in some detail the role of GIS in developing comprehensive strategic plans, suggesting that the type of information provided by a GIS makes it uniquely capable of enhancing the planning process for a variety of public and private organizations.

Because of the increasingly complex nature of the environment within which organizations operate, today's managers face a variety of new and challenging demands. Novel and more complicated problems have given rise to a better educated workforce using new technologies to help them manage their professional responsibilities. The introduction of computer technologies and other systems-integration tools has helped to increase the speed of communications, thereby quickening the pace of day-to-day activities. Similarly, the world outside the organization has become more complex and faster paced. Issues that might once have been viewed from a narrow parochial perspective are now routinely addressed as part of a wide global view.

As the typical organization's environment—both internal and external—has become more complex, the organization's decision making has become more challenging. Strategic planning (sometimes referred to as "strategic management") developed as a management tool over the 1970s and 1980s to meet the challenges posed by the increasingly complex environment of an organization. For a number of public and private organizations, a GIS is a uniquely qualified tool to aid in long-range planning and objective setting. For these organizations, geographic information is a vital element in their planning process. Whether the organization is a small local government agency concerned with new residential development and subsequent questions of rezoning and infrastructure expansion or is a huge international paper company intent on ensuring a supply of timber for its long-term operations, geographic information of the type provided by a GIS can serve as a key element in setting a strategic focus for these organizations.

Because GIS is often central to the process of strategic planning, we focus in this chapter on the relation between an organization's planning process and the role that a GIS can play in these activities. We develop, in some detail, the basic elements in a strategic plan. Furthermore, we address a generic model of strategic planning to show the interrelatedness among the various organizational and human factors that can influence the planning process. Finally, we discuss the specific stages in the planning process and the reasons why geographic information of the type provided by a GIS can be so beneficial.

What Is Strategic Planning?

One of the earliest comprehensive and relevant definitions of *planning* was offered by Scott (1963: 4), who stated that "planning is an analytical process which involves the assessment of the future, the determination of desired objectives in the context of that future, the development of alternative courses of action to achieve such objectives and the selection of a course (or courses) of action from among those alternatives." Scott's definition is important because it identifies several key planning elements and activities. These significant points in his definition need to be underscored because they cut to the heart of any approach to strategic planning. First, planning is a *process*—that is, it represents a dynamic, ongoing effort on the part of the organization and its members. Planning is not a static event. It does not simply occur at well-defined intervals, is engaged in for a prescribed time period, and is then abandoned. Effective strategic planning is a robust and continuous activity on the part of the organization as it acknowledges that change is unpredictable, continual, and potentially highly significant to the organization's future operations. Furthermore, because it is analytical, the planning process is essentially rational in its intent and its efforts.

The second important element in the definition is the idea that planning is *future-directed*, with the intent focused on determining the appropriate objectives for an organization within the context of possible significant changes in the organization's operating environment. Third, strategic planning is *flexible*. It requires planners to anticipate a variety of possible scenarios, to develop alternative courses of action to deal with these scenarios, and—depending on the environmental threats and opportunities encountered—to select the appropriate course of action that can most effectively address the organization's concerns.

The major innovation of strategic planning is its underlying assumption that the world is uncertain and unpredictable. An organization must understand its uncertain environment (both internal and external) if it is to be successful in accomplishing its mission. Understanding the environment is an ongoing task since the environment is subject to change; hence, the genesis of the phrase *strategic management*. Figure 5.1 illustrates the various components of the strategic planning process. This figure is important because it illustrates the interrelatedness of the component elements and characteristics of the planning process. These elements include (1) the organization's *analysis process*, (2) its *expected future*, (3) its *perceived competences*, (4) its *top managers' preferences and values*, and (5) its *stakeholders' priorities and power*, all of which affect (6) the organization's *strategic plan*.

The Analysis Process

The *analysis process* refers to the formal methods by which the organization engages in strategic planning—that is, the information that staff members deem relevant to their operations and that they choose to collect, the specific steps that they take in developing their strategies, and the analytical process based on hard data (e.g., computer-provided analyses) or generally consisting of "soft" data based on word-of-mouth or other qualitative information.

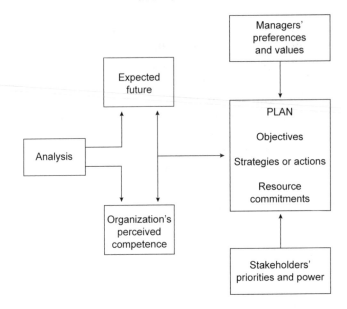

FIGURE 5.1. Steps in a strategic planning process. From Camillus (1986). Copyright 1986 by Lexington Books. Reprinted by permission.

The Expected Future

The *expected future* of an organization simply means that the organization develops a variety of scenarios to assess as many possible future states as can be envisioned by the planners. Essentially, the planning staff are asking a series of "What if?" questions in order to cover every possible future contingency. As an example, one of the contributing events that led many companies to develop their own strategic planning departments was the series of OPEC-generated "oil shocks" of the 1970s. So many companies were seriously hurt by these embargoes and price hikes that they determined never to be caught unaware in the future. Although the *expected future* means just that (the future that seems most probable), it is also important to develop several alternative scenarios and analyze appropriate responses of organizations in order to reasonably ensure the likelihood of not being surprised by future events.

Perceived Competences

In tandem with the expected future, strategic planning needs to acknowledge an organization's set of *perceived competences*. What is it that a particular organization does well? If the firm is private, the question that needs to be asked is, What is this firm's competitive advantage over other companies? What has the firm done, and what does it continue to do, that makes it successful relative to its competition? If the organization is public, such as an agency of a local or county government, it still must examine its activities to determine in what areas it possesses a set of distinctive competences. For example, some towns pride themselves on the recreational elements of their community (water sports, skiing, golf, etc.). Other towns may take justifiable pride in their low crime and public safety records. The logic behind asking "What does an organization do well?" is to establish a framework for planning. Sound strategic plans almost always are intended to build upon current strengths rather than to reinforce inadequacies.

Top Managers' Preferences and Values

Obviously, a serious contributing factor to the focus of a strategic plan will involve *the preferences and vision of the organization's chief administrators*. Top management has the ultimate responsibility for the welfare of the organization. Because of its power, top management can have a tremendous effect on the future direction of the organization. In fact, one of the long-acknowledged duties of senior administration is to set long-term goals and strategies for organizations. As a result, it is easy to understand how the values and beliefs of top management will have a great influence on the types of plans and strategic directions taken by an organization.

Stakeholders' Priorities and Power

A *stakeholder* is defined as any group or individual with a potential interest in or effect on an organization. In other words, anyone who has a "stake" in what the organization proposes to do is referred to as a "stakeholder." Stakeholders may have a great deal of power in some organizations (e.g., a large stockholder of a publicly held company), while being relatively powerless in others. One question that it is necessary to ask prior to committing to a strategic direction and developing subsequent long-term plans is the nature of the organization's stakeholders. Are they powerful? Must they be consulted before strategic planning takes place? To what degree are they capable of altering strategic plans and, in effect, redefining the organization's mission?

An organization must assess the influence of its major stakeholders on a continual basis to ensure that they (the stakeholders) are kept apprised of the directions that the organization takes. Within public organizations, many towns hold town meetings or regular meetings of their city councils as a forum for stakeholder groups to present their ideas, to register their support for or objections to strategic choices, and to attempt to influence public policy. For the public sector, stakeholder groups may take on many forms, including small, localized special interest groups as well as large nationally based efforts (e.g., Mothers Against Drunk Driving, the National Rifle Association). The groups work to influence public policy at all levels and are to varying degrees often successful in setting strategic agendas and influencing long-term planning.

An Organization's Strategic Plan

The final element in the generic planning process is the actual development of the *strategic plan*, which is subject, as we have noted, to the myriad groups and individuals who are to varying degrees capable of influencing these plans. As you can see from Figure 5.1, there are three basic elements to the creation of a strategic plan. First, the organization develops a set of objectives or goals for the future. These goals may include short-term goals (milestones to occur in less than 1 year), medium-term goals (to occur in 1–3 years), and long-term goals (to occur 5 years to well over 20 years in the future). The objectives signal the planning team's analysis of the likeliest future directions for the organization, based on their interpretation of current data and trends. Following the establishment of these objectives, a series of specific strategies or steps are agreed to. These strategies consist of the activities that the organization believes are necessary to achieve long-term goals. For example, if one goal for a small seaside town is to promote additional tourism and development over the next 10 years, one intermediate strategy necessary for realizing this objective might be to rezone oceanfront property for multiple dwellings and begin infrastructure expansion and dockside renovation.

Finally, the third important step in the strategic plan is an organization's commitment of resources to aid materially in supporting the strategies necessary to achieve long-term goals. Programs and strategies that are not materially supported by top management are almost invariably doomed to failure. Providing resources in support of strategic choices is often a difficult prospect, particularly in times of tight budgets and shrinking revenue. Yet difficult times are also often the test of an organization's commitment to its strategic objectives. When difficulties set in, the easiest avenue to follow is to revert back to old familiar patterns, including outmoded plans and strategies. This approach, while often convenient in the short run, is almost always disastrous for the long-term viability of an organization. For example, in the 1980s Ford Motor Company suffered from flat sales and customer dissatisfaction due to a series of poor vehicle design decisions and low overall vehicle quality. After spending the better part of two decades in revamping designs and adopting a more "customer-focused" approach to business, Ford is again stumbling in the current decade, as it began to make again the same design and quality errors that had gotten it into trouble years earlier. One critical miscalculation has been too heavy reliance on large, gas-guzzling cars and trucks that are increasingly unattractive because gas prices are pushing $3 a gallon in the United States. Thus, Ford demonstrates again that, while often alluring, the short-term "quick fix" is usually a serious long-term disability for an organization.

Based on the interaction of each of the elements in strategic planning, we begin to see the emergence of a complex picture in which a variety of factors can influence the development of an organization's objectives and strategies. The equation is made even more complex by the existence of a number of stakeholder groups, each of which has a strong vested interest in influencing the development of an emergent strategy. The model shown in Figure 5.1 offers some indications of the complexity of the true strategic planning process. However, while complex, this model does not offer a comprehensive outline of the actual steps that managers need to take in order to develop a viable strategic plan. The following sections of this chapter discuss the main elements of a strategic plan—in particular by demonstrating the role of geographic information in creating long-term goals and the strategies to realize those goals.

The Process of Strategic Planning

While the potential payoffs from effective planning can be tremendous, the strategic planning process itself poses a series of important challenges for managers. Generally defined as *the process of determining an organization's long-term goals and then identifying the best approach to attaining those goals*, strategic planning represents a critical (some would argue *the* critical) activity in maintaining the current viability of the firm as well as positioning it for future success. In effect, we use strategic planning as a means to organize the present on the basis of our projections about the future. That is the reason that the design of a strategic planning process is so important. As Camillus (1986: 59) has noted, the "design of the process poses a particularly significant challenge in that it is an important means by which rational, economic analyses can be integrated effectively with behavioral, political realities." This statement underscores a key aspect of the planning process: that it does not consist entirely of the use of analytic methods to arrive at the best (profit-maximizing) alternatives and objectives. Rather, strategic planning recognizes the inherent interaction of economics with political and human realities. Consequently, in engaging in the strategic planning process, three important questions must be addressed prior to active information search and objective setting.

1. *What* should be analyzed? This is the rational/economic perspective that seeks to determine what data are important for developing a strategic plan.
2. *Who* should conduct the analysis? The political realities of most organizations suggest that certain individuals have a greater stake in long-term objectives—and consequently their decisions will carry greater weight in strategy development and goal setting.
3. *When* and how often should analyses be conducted? The timing perspective of strategic planning is also crucial. What should be the strategic planning cycle for our particular organization?

Bases for Strategic Analysis

The typical strategic planning process involves an in-depth scan of both internal and external elements of an organization. Three generally accepted dimensions must be addressed through this scanning process: *the industry situation, the competitive situation,* and an internal analysis involving *the organization's own situation* (Thompson & Strickland, 1987). Each of these dimensions is addressed below. It is important to point out up front, however, that the strategic situation faced by private and public organizations is clearly different relative to industrial and competitive forces. Certainly for private (profit-seeking) firms, an in-depth market and competitive analysis is essential for long-term survival. On the other hand, municipalities and other government agencies are not, by their very nature, threatened with the same type of competition and concern for profit maximization. As a result, readers must carefully consider each element in the strategic scanning process to determine its particular applicability to their own organization's situation. Some of these elements may simply not be applicable to all types of organizations. Nevertheless, this framework offers an important starting point for strategic planning by emphasizing some of the key aspects of an organization's environment, both internal and external. Furthermore, it will be shown that geographic information of the type generated by many GISs can be central to developing a better overall understanding of the environment within which an organization is operating.

Industry Situation Analysis

The industry situation analysis involves an examination of overall industry structure, direction, economics, and its long-term viability (attractiveness). Obviously, for public agencies and municipalities, issues of economics and attractiveness take on a different meaning than what they have for private firms. Nevertheless, the basic elements of a situation analysis are equally important and have meaning within the public sector. *Industry* can be broadly defined as the set of organizations that are so similar that they are drawn into competition with each other for the same client base. The underlying purpose of the industry situation analysis is to gain a sense of the long-term potential for effectively competing within an industry. As a result, the industry situation analysis requires planners to take a clear look at their own organization relative to expected and potential changes within the overall industry or economy. Based on the identified characteristics of an organization and its external environment, is it more or less likely to remain a strong competitor into the future?

Table 5.1 gives some examples of the types of questions that planners need to ask in conducting an industry situation analysis. That is to say, in asking how the industry is structured, planners attempt to determine whether the industry is relatively closed or if it can be readily entered. For example, the forestry and paper products industry may be thought of as relatively closed in that it has high barriers to entry. To become a viable competitor, a company must not only engage in a major capital outlay for plant and equipment (sawmills, transportation, etc.), but must also invest heavily in land purchases to ensure a steady supply of the timber that serves as the principal raw material. Because of these high barriers to entry, a central question that must be resolved is the degree of attractiveness that the industry still holds in spite of high start-up costs.

TABLE 5.1. Important Questions in the Strategic Planning Process

Industry situation analysis

How is the industry structured? Are barriers to entry high or low?
What general trends or driving forces can be discerned?
What are the key success factors in this industry?
What long-term strategic issues does this industry face?
Should we be in this industry? Why or why not?

The competitive situation

What competitive forces exist? How strong are they?
What do we know about key rivals? Their strategies and competitive strengths?
Where will future competition come from? What will it be like?

The firm's own situation

How well is our present strategy working? Will it require minor adjustments
    or wholesale changes?
What does our SWOT analysis tell us? What are we doing right and, more
    importantly, what are we doing wrong and how can we fix it?
What is our competitive advantage?
What specific strategic issues do we need to address?

Some examples of other things that must be considered in performing an industry analysis include an identification of key success factors in the industry, the long-term trends the industry faces, and a general question of whether the firm should even continue to operate in the industry. Again, to use the example of the forestry company, one potential long-term trend that could bode ill for firms of this type is the enhanced concern for protecting natural resources promoted by vocal advocacy groups such as the Audubon Society and Greenpeace. The power of these groups poses a serious threat to the long-term profitability of natural resource–based companies. To illustrate, the continued exploitation of the Pacific Northwest's timberlands has been severely restricted as a result of the threat to the spotted owl. The federal government has at present made the determination that preserving this endangered species outweighs the potential loss of thousands of forestry jobs.

Public organizations are also in a position to conduct their own version of an industry situation analysis. While they do not generally face the same types of pressures toward profitability, they are just as concerned about the long-term survival of their agencies or municipalities. Consider, for example, the case of the March of Dimes, which was established originally as a charity to help those afflicted with childhood polio. In the wake of Dr. Salk's polio vaccine, which essentially eliminated the threat of polio, the March of Dimes was forced to perform an industry situation analysis to determine where its efforts would now be needed. As a result of this analysis, the March of Dimes chose to involve itself with birth defects. Public organizations and governments need to engage in industry scanning and analysis for their long-term survival to the same degree that private firms do. The difference, as illustrated in Table 5.1, lies in the fact that the type of questions that public organizations need to concern themselves with are often markedly different from those of private firms.

The Competitive Situation

While the industry situation analysis forms a backdrop for in-depth examination of an industry's competitive posture—sources of competition, barriers to entry, key industry success factors—the competitive analysis helps organizations to engage in a more focused search for advantage vis-à-vis their direct competition. In analyzing the competitive situation, three lines of questioning are key:

1. What is the nature and strength of the various competitive forces?
2. What are the competitive positions and strengths of key rivals? What are their strategies? What do they do better than we do?
3. What can we expect our rivals to do next?

The importance of the competitive analysis segment of the overall strategic planning process cannot be overestimated. It is only by conducting an analysis of its competitors that an organization comes to a better understanding of its own position in the industry. In other words, it is through an organization's efforts to gain an understanding of the strengths and weaknesses of its competition that it is able to gain a greater understanding of its own strengths and weaknesses. Even firms that are in attractive industries may find themselves unprofitable because they have allowed themselves to be placed in a weak competitive position against aggressive rivals. Hubris, poor planning, and willful blindness can all cause an organization to downplay competitive position analysis that will almost inevitably result in a severely curtailed market. To illustrate: Consider the real example of a city with a harness-racing track that had traditionally been the source of enhanced revenue for the municipality. Under a series of misguided assumptions that (1) casinos will never work, (2) casinos and race tracks are nonsubstitutable forms of entertainment, (3) that "real money" can only be made in racing, and so on, the city council chose to ignore reports of a neighboring community's efforts to construct legalized gambling casinos. The neighboring community's casinos became so popular that they severely lowered racing revenues to the point where the track had to be closed. While this is one example, in other cases cities have suffered a real loss of revenue through a competitive force that they chose not to take seriously until it was too late.

The above example highlights an important point that must be stressed about the benefits of competitive analysis: namely, its relevance to public administration organizations. The point can justifiably be made that a town planning board or city agency is not threatened by the actions of "rivals" in the same manner or to the same degree as private or not-for-profit organizations are. If public administration GIS managers do not face the same sorts of competitive pressures, what is the benefit to them in conducting competitive situation analysis? In order to answer this question, it is important first to understand that "competition" arises from a number of sources and in relation to a number of functions—other than simple profit making, as engaged in by privately held firms. The above example of a town using harness racing as a form of municipal revenue generation is a case in point. Should another municipality within a convenient distance engage in similar activities, their actions could pose a competitive threat to the town that has promoted racing. Likewise, one community's decision to raise property taxes to a higher level than those of other neighboring communities could lose considerable support if residents perceived no difference in municipal services offered, and hence relocated to areas with lower taxes. This is consistent with the Tiebout (1956) hypothesis, which suggests that when local governments provide a variety of public goods and services, consumers "vote with their feet" by moving to the towns that provide their preferred mix of goods and services.

The point that must be emphasized is that communities and other types of public administration may not face the same pressures from competition that has become synonymous with American business. This does not mean, however, that they do not face their own forms of competition from neighboring communities as well as from other public agencies at state and federal levels. As a result, it is important that both private- and public-sector GIS managers consider the potential effect of competitors on their operations and, consequently, engage—to a varying degree of sophistication—in some form of competitive analysis.

Analyzing the Competitive Situation: The Five-Forces Model

An extremely useful tool for gaining a better understanding of the nature of competitive pressures is through the five-forces model developed by Michael Porter (1979) of the Harvard Business School. Porter argues that any organization must be aware of five distinct forms of competitive pressure in the marketplace:

1. *The jockeying for position among rival organizations in search of competitive advantage.* Each firm attempts to employ its own brand of competitive strategy in order to gain a favorable position relative to its chief rivals. This "competitive edge" becomes the key to long-term success by maintaining an advantage over other rival organizations.

2. *The intrusions and threats from the substitute products of organizations in other industries.* Competitive forces can also arise from the threat of substitute products from other industries. For example, if soft drink manufacturers price their products too high, they run the serious risk of losing market share to fruit juice or powdered drink producers.

3. *The potential for entry into the industry by new competitors.* If the barriers to entry into a market are perceived to be low, the industry's attractiveness could increase, encouraging new competitors to enter the market. Furthermore, this threat of new entrants can have a constraining effect on firms already in the marketplace. The airline industry offers a good example of the need for constraining influences. Owing, in part, to the federal government's perceptions of inequities in fare structures and monopolies within certain markets, it deregulated the airline industry in the late 1970s. Deregulation led to a large number of new firms entering the airline business in the late 1970s and early 1980s. In order to gain a competitive advantage, airlines engaged in a series of fare wars, which initially had the effect of shaking out a number of small and less financially sound airlines. However, deregulation also established a threshold model for airfare pricing that has severely limited upside price rises, even as airlines face huge new expenses in the form of energy prices and labor contracts. Thus, new entrants into the airline industry (e.g., Jet Blue or Southwestern) have had to find a unique niche in either pricing or service that allows them access to the traveling public.

4. *The power and bargaining leverage of suppliers.* Important suppliers of all types of raw materials for an organization can influence not only that firm's ability to conduct business but also the strategic posture adopted by the company. The makers of semiconductors have such an influence on the operations of calculator and computer manufacturers that at one time IBM purchased a significant share of Intel Corporation's stock (Intel is the world's largest manufacturer of semiconductor chips). It is also important to note that a supplier's bargaining position and power over an organization often lies in direct proportion to the lack of substitutes for the supplier's product. For example, the power of suppliers of glass bottles to soft drink manufacturers is constrained by the latters' ability to use aluminum and plastic containers as substitutes for glass.

5. *The power and bargaining leverage of customers.* Just as powerful suppliers can influence the operations of firms within an industry, so too can powerful customers. When customers are few in number and are actively being pursued by a number of rival firms, they have a great deal of power in negotiating advantageous deals. Furthermore, there are many cases in which a firm has signed a series of exclusive deals with suppliers for their products. Because the one company now poses as an exclusive customer, it can shape the future direction of the company for which it is the chief customer.

Figure 5.2 illustrates Porter's five-forces model. One point to note is the tremendous degree of interaction that these forces often have on each other as well as on the organization. In conducting a competitor analysis, the organization must be aware of the nature of each of these threats and the related fact that energy spent attending to one threat may actually encourage another threat to appear. As an example, let us assume that XYZ Corporation perceives that its chief rival has a technological advantage over it through more sophisticated manufacturing procedures. In an effort to reduce that technological advantage, XYZ Corporation develops a new low-cost procedure that allows it to be competitive with this chief rival. The advantage is that the company has negated the technological edge in manufacturing that the rival held. However, because of the new low-cost breakthrough, the first company has, in effect, lowered a major barrier to entry into the market and to new competition and substitute products. As a result, owing to the interaction of the five forces of competitive pressure, the first company may now actually be worse off than it was before.

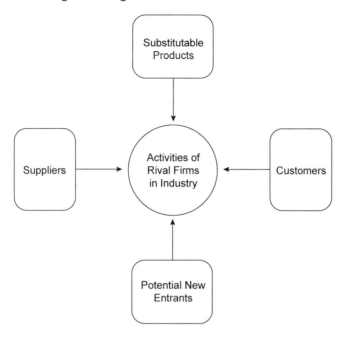

FIGURE 5.2. The five-forces model of competition. Reprinted by permission of *Harvard Business Review*. From "How Competitive Forces Shape Strategy" by M. E. Porter, March/April 1979, 57(2). Copyright 1979 by Harvard Business School Publishing Corporation; all rights reserved.

An Organization's Own Situation

In addition to conducting a detailed external analysis of both the nature of the marketplace within the industry and the competitive stance of the organization relative to its competitors, it is extremely important that the organization devote sufficient time to performing an internal analysis. Up to this point, the focus of the organization's analysis has been on the external arena, assessing the industry as a whole as well as the firm's competitive posture within the industry. At this stage, the spotlight shifts to the organization itself. Simply put, an internal situation analysis requires the organization's planners and top management to conduct a detailed assessment of the current state of their organization and to address some compelling questions. How well is the present strategy working? What do we do well? In what areas do we need improvement?

The primary methodology that drives the internal analysis goes by the acronym SWOT, which refers to the organization's assessment of its internal *strengths* and *weaknesses* as well as its external *opportunities* and *threats*. First and foremost a SWOT analysis consists of an honest appraisal of the current state of the organization, one that can be extremely useful for sizing up the company's activities and operations. The underlying concern is that the analysis be conducted for the purpose not of validating the conduct of the organization, but of analyzing the company's *strengths* and *weaknesses* as candidly as possible. A SWOT analysis is not intended to be a self-congratulatory process, but is intended for reaffirmation. It is a highly useful method for staying abreast of an organization's activities.

Table 5.2 gives some examples of the types of issues that are important in performing a SWOT analysis. The firm's answer to these questions will go far toward pointing the organization in the direction of remedial activities to correct its defects and to solidify its strengths. It is important that, in answering these questions, as wide a range of involved managers as possible take part in the process. The obvious advantage of using multiple respondents to assess an organization's current state (i.e., its strengths and weaknesses) is that some concerned parties or individuals, particu-

TABLE 5.2. Examples of Key Questions in the SWOT Analysis

Potential strengths?

Distinctive competences?
Sufficient financial, human, and raw
    material resources?
Technological innovativeness?
Manufacturing efficiency?
Good management?
Cost advantages?
Industry leadership?
Good functional integration?
Comprehensive information system
    (including GIS)?

Potential opportunities?

New markets or niches to enter?
Diversification into new product lines?
Decline of competitor's position?
New customer expectations?
Development of acquisition strategies?
Vertical integration?
Advantageous demographic changes?
Potential for market or resource
    development

Potential weaknesses?

No clear direction?
Falling behind the experience curve in
    research and development?
Narrow product line?
Obsolete or inefficient production?
High cost of operations?
Poor worker–management relations?
Inadequate functional skills?

Potential threats?

Likely entry of new competitors?
Changing customer needs or tastes?
Encroachment by substitute products?
Enhanced bargaining position of
    suppliers?
Enhanced bargaining position of
    customers?
Adverse demographic changes?
Market growth slowing, stagnant, or
    declining

larly from the top management group, may be ignorant of or willfully blind to weaknesses within company operations.

Organizations have differed in their approach to conducting and using a SWOT analysis. Some firms have compiled detailed checklists similar to, but more comprehensive than, that shown in Table 5.2. Following the creation of these checklists, the company's managers perform their SWOT assessment using some form of scale (e.g., a 5-point scale where 1 = "Needs improvement" and 5 = "Outstanding"). The strategic planning team then conducts an analysis of how their own organization stacks up against the checklist and performs some "best guess" SWOT analyses on their chief rivals. The use of SWOT analysis in this manner has some advantages in that it allows for comparisons between an organization and its competition. The company can make some tentative determinations regarding the way in which it stacks up in the industry. While this is one approach to SWOT analysis, the important point to remember is that no matter how the analysis is conducted, the benefits to be derived from actually engaging in the process can be extremely eye-opening, particularly for firms that have never made this type of systematic assessment.

Strategic Planning and GIS

What, then, is the role of geographic information in the strategic planning process? The obvious response to this question is that for many organizations GIS have become intimately tied to the entire strategic planning process. In effect, they become an important element in this process. One of the points we have tried to make throughout this chapter is that the strategic management process is intensely information-dependent. In order to engage in sufficient industry and competitor analysis, a great deal of both internally and externally generated information is required. One of the generally accepted strengths of a GIS is its ability to collect and sort the myriad pieces of information, thus assisting planners tactically (concerning day-to-day decision making) as well as strategically. The strategic use of a GIS is seen every day in both public and private organizations. For example, urban planning agencies—tasked with the responsibility of developing comprehensive and sound policies for zoning, urban renewal and development, land reclamation, and so forth—have found that GIS offers an efficient and effective technology for providing the type of information so important in the decision-making process. The GIS has become a central storage facility for the thousands and thousands of diverse pieces of information that can be stored against the future possibility of their being needed in making strategic planning decisions.

The role of the GIS in strategic planning becomes even more apparent when we reexamine the various elements of the strategic planning process set forth in this chapter. We have suggested that the three key questions first needed to be addressed are: *What* should be analyzed? *Who* should conduct the analysis? *When* and how often should the analysis be performed? We can see that the use of a GIS gives the planning group tremendous flexibility in addressing each of these crucial questions. First, if an organization has taken adequate time to ensure that it is collecting and storing all necessary information for its information system, the question of what should be analyzed becomes much easier to deal with than in other situations in which the organization does not have any form of comprehensive information system. In the latter case, the company may be forced to make decisions based on insufficient information. Because the company has not taken the steps to ensure an adequate information storage and retrieval system, the whole question of what should be analyzed becomes somewhat academic. Such an organization is forced to analyze whatever information it has available, the result of which is often making "seat-of-the-pants" decisions based on past experience or gut instinct rather than on analysis of sufficient information.

Another benefit of the GIS refers to the question, Who should be conducting the strategic analysis? Most existing research points to the fact that political realities often place certain higher level individuals in the position of exerting more influence over the decision process. While it is true that higher level individuals can often influence the nature of goal setting and information search, the availability of a GIS has something of an egalitarian effect on the organization. In other words, through the creation and use of a GIS, information becomes more accessible to a wider range of people. Rather than making its decisions in a "black box" manner, in which the majority of the organization is unaware of how decisions are made, an organization that uses a GIS puts the sources of information literally at an individual's fingertips. As a result, a wider range of people can make use of the information that is necessary to engage in effective decision making. Furthermore, a GIS can serve as a check on individuals misusing the power of their position. When the information on which a decision was based is instantly available to a number of people, there is a greater possibility that these individuals will challenge the basic assumptions of the decision maker.

Finally, a GIS also answers the questions of when and how often analyses should be performed because it offers strategic planners the capacity for continual planning. Because a major information source is always available and instantly accessible, the planning process has the capacity to operate continually. As new data are fed into the storage system, information is constantly upgraded. These upgrades are intended to maintain the effectiveness of a major planning tool. As a result, strategic planning has the capacity of being performed on a continual basis, with the only limitations being the amount of new information fed into the system, the frequency with which the system's database is upgraded, and the availability of trained personnel within the organization to access this information and make use of it in the planning process.

As a final point about the role of the GIS in the strategic planning process, we can see that geographic information is useful for conducting both a situation analysis and an internal analysis of an organization. The industry analysis requires that an organization make a series of determinations about the nature of its external relationship within the industry. How will the organization compete? Is the industry still as attractive to the organization as it once was? Some forms of geographic information provide important answers to questions about industry status. For example, a hydroelectric facility may determine from census data and water flow analysis both that its client base is shrinking and that, because of recent engineering projects, water levels are falling to the point where it will become increasingly costly to continue generating electrical power. These determinations could signal that the industry is rapidly losing its attractiveness, which could thus lead to significant strategic reorientation.

Furthermore, the GIS can provide important information as part of an organization's internal SWOT analysis. Through the analysis of geographic information it may be found that there are significant threats to a city's infrastructure from old systems or neglected maintenance. As in the case of Chicago's recent downtown flooding, information had been available to suggest that there was imminent danger of catastrophic flooding, which was generally disregarded until too late. New York, as well, is currently facing similar problems with its infrastructure (most notably, bridges). The GIS can provide not only important information for municipalities regarding future GISs and opportunities in the areas of development but also warnings of potential threats from a number of sources.

Conclusions

In this chapter we have examined some of the key elements in the strategic planning process for both public and private organizations and the central role that a GIS can play in this process. Strategic planning has become increasingly important for most modern organizations within both the public and the private sectors. It gives these organizations a "sense-making" device that can help managers understand the relationship of their organization to the external environment (including competitors). Furthermore, it offers these organizations the opportunity to maintain a proactive, rather than a reactive, relationship with their environment. There is an old saying, "Those who fail to plan, plan to fail." The central thrust of this saying for most organizations is that strategic planning is no longer a luxury but a necessary part of their activities. For most public and many private organizations, geographic information of the type provided by the GIS is a powerful tool in the planning process. This chapter has attempted to explicitly link the process of strategic planning with the functions of the GIS, demonstrating the essential and growing interrelatedness of their operations.

Chapter 6

Implementing a GIS
THEORIES AND PRACTICE

A necessary part of the evolution of ever more complex IS has been the introduction of these systems into organizations to aid in their day-to-day operations. One of the long-standing difficulties with the development of enhanced IS capabilities within organizations has been the problem of gaining widespread acceptance and use of these systems. This implementation problem has continued to inhibit the successful implementation of IS technologies throughout their history. It is our contention that the primary problems that underlie most implementation concerns are usually organizational rather than technical. That is to say, in contrast to technical problems accompanying the introduction of new innovation or IS technologies (which are usually quickly surmounted), organizational problems of implementation refer to the human aspects that can inhibit or limit the potential acceptance and use of new system innovations. A considerable body of research evidence has shown that paying attention to the behavioral or human factors within the organization can greatly facilitate the likelihood of acceptance and use of new technological innovations.

Numerous studies have addressed the adoption and diffusion of technological innovations in other fields (e.g., Downs & Mohr, 1976; Rogers, 1962; Schultz & Slevin, 1975). However, few formal social science investigations of geographic information technologies have been reported (Onsrud & Pinto, 1992; Wellar, 1988a, 1988b; Wiggins & French, 1991). It

Portions of this chapter are from Pinto and Onsrud (1997). Copyright 1997 by Taylor & Francis. Adapted by permission.

is important to note that the terms "diffusion" and "implementation" are used interchangeably within this section. Historically, the concepts of diffusion and implementation were independently derived from different sources. "Diffusion" usually referred to the acceptance and use by some subset of the general population of scientific or technological innovations. In contrast, "implementation" typically referred to the acceptance within organizations of new technical processes or models. However, over the last several years, the two terms have come to encompass the same idea. Therefore, for the purposes of this chapter, diffusion and implementation are both intended to refer to the *process through which an innovation is communicated through certain channels over time among the members of a social system* (Rogers, 1983).

An understanding of the diffusion process can help those who could benefit from an innovation, such as a new technology, to begin accruing those benefits earlier. By identifying crucial social factors and processes in the adoption, implementation, and utilization of a technology, we would expect to predict the decision-making responses of individuals, groups, and organizations more accurately, and therefore to accommodate or redirect these processes through prescriptive strategies. Since they identify crucial human and technical factors within classes of potential users, diffusion studies also have the potential for directing the design efforts of system developers to those system characteristics and improvements most valued by end users.

In this section we explore some of the important issues in the diffusion and implementation of new IS innovations, examining the historical roots of the implementation problem, some of the approaches to implementation that have been examined to date, and the nature of successful implementation. We suggest, through the development of a framework model, that implementation success is a multifaceted issue, comprising a number of diverse but equally important criteria.

The concept of implementation in the context of organizations may be viewed as a "change phenomenon," or a process for creating organizational change. Initially, the problem of implementation was discussed in the context of frequently ineffectual attempts on the part of operations researchers and management scientists to generate enthusiasm for and use of a myriad of new organizational innovations intended for use by practicing managers. More recently, the problem of implementation has been represented as the frequent failure to create some degree of desired organizational change through the introduction of a new information system, program, or model. Indeed, Schultz, Ginzberg, and Lucas (1983) define implementation success in terms of changed behavior on the part of organizational members. Implementation, then, in its most basic sense, has increasingly come to be defined as a new IS, program, or model that has been accepted by organizational personnel, the results of which system change the decision-making processes of the personnel (Schultz et al., 1983; Schultz & Slevin, 1975).

Literature in the area of implementation, though significantly increasing in volume in recent years, remains largely unfocused. That is to say, many writers in the field have discussed implementation in a general sense, paying little attention to the type of implementation effort being performed, perhaps in the belief that there exists little or no real difference between various forms of organizational implementation (information systems, projects, OR/MS [operations research/management science] models, etc.). Bean, Neal, Radnor, and Tansik (1975) discussed the implementation of OR/MS as seemingly interchangeable systems and projects. Furthermore, Harvey (1970) developed a set of critical success factors in the implementation process without regard to distinguishing between types of implementation.

Early work on implementation tended to focus on the important actors in the implementation process rather than on the type of implementation being considered. For example, Churchman and Schainblatt (1965) discussed cognitive styles and the need for mutual understanding between the researcher and the user as a way of improving the chances for implementation success. Schultz and Slevin (1975) argued the need for formal feedback channels to encourage constant communication between researcher and user, in the belief that a key focus of implementation research should be on the organizational actors.

The topic of implementation has also received a great deal of attention in the public policy and public agency literature. Pressman and Wildavsky (1973), who coauthored one of the earliest works on implementation, discussed the implementation problem from both theoretical and practical perspectives and used the public policy forum as a basis for their work. Furthermore, Majone and Wildavsky (1978) discussed the concept of implementation as a controlling and interacting variable, distinguishing between activities necessary for public policy enactment and those related to the implementation of previously developed policies. While much of the public policy work on implementation has largely remained unintegrated with the rest of the field of organizational research, the work of Wildavsky and others demonstrates the strong connection that implementation has with both public- and private-sector problems.

Definition of Implementation Success

Schultz and Slevin (1975) were correct in their early assessment of implementation success when they posited that any successful implementation effort is predicated on the technical efficacy of the system about to be introduced. In other words, the technical viability and workability of the GIS represents the first necessary but not sufficient condition for that system's successful adoption. Prospective users of any GIS must first be confident that the technology they are considering does, in fact, work, that is, that it accomplishes the tasks that its advocates claim it can perform. What this discussion is referring to is the notion of "information system quality," as identified by DeLone and McLean (1992). Some of the more obvious measures of system quality, particularly within the context of GIS, would include system response time, ease of online use (user friendliness), and reliability of the computerized system (absence of consistent downtime). These all represent some of the more common and well-accepted determinants of a GIS's technical quality and should rightfully be addressed in assessing the chances for a successful introduction. Some technical characteristics are easily comparable across alternative GISs, such as built-in features, expandability, system speed, and so on. Other aspects of system quality (e.g., user friendliness) may be more qualitative and difficult to rate, let alone compare, with GIS alternatives. Certainly, given the nervousness regarding computerization felt by many members of organizations, attempting to develop and introduce a non-user-friendly system may be a difficult and ultimately futile process.

A second measure of the technical viability of the GIS is the idea of information quality. The importance of information quality derives from the notion that any system is only as good as the information it delivers. In other words, rather than simply considering how "good" an IS is (system quality), a better representation of quality would examine the outcomes of the GIS, that is, the accuracy of the data. In addition to data accuracy, some of the most important metrics for assessing the information quality delivered by a GIS are proposed data currency, turnaround time, completeness of the data produced, system flexibility, and ease of use among potential clients of the system. Additional elements of information quality are ease of interpretation, reliability, and convenience. These criteria are all examples of ways in which we can rate the quality of the information generated by the GIS and, as in the case of system quality, they provide a context for the willing acceptance and use of the system by its intended target departments. It is important to note, however, that, as with any information storage and retrieval system, the quality of information produced lies in direct proportion to the quality of the information inputted. Consequently, an additional implication of this model is the need to enforce quality control throughout the organization, upstream of the GIS. A "quality" GIS cannot obviate the work of poor quality inputting.

The third aspect of GIS success is the importance of information use. "Information use" refers to the obvious point that information of any sort is only good if it is accepted by organizational members and used in their decision-making processes—that is, the information needs to be consumed by its recipients. Underscoring the difficulty of gaining system acceptance and information use is the problem of attempting to change employee behavior. During a series of interviews the authors conducted with representatives from municipal governments in some northeastern states, we were shown the town planning office's GIS, a PC-based version of a popular GIS product. The system was quite literally gathering dust as the members of the department continued to make use of paper maps and old charts for zoning and public works decisions. When we asked the planner why his subordinates were not performing these routine tasks with the town's GIS, he replied that the PC's monitor had broken down over 6 months ago, but that suited everyone just fine as they had never had much use for the GIS. Here was a clear-cut case of the problem with acceptance and use of a system. So disinterested were the planners in the GIS-created data that they had seized on the excuse of a broken PC as a basis for continuing with their old practices.

Beyond the obvious point that a successfully implemented GIS must be used is the more subtle issue of the different levels of use that may be found. For example, one study of IS implementation identified three different types of use:

1. Use of a system that results in management action.
2. Use that creates or leads to organizational change (different ways of performing standard tasks).
3. Recurring use of the system (Ginzberg, 1978).

Another study expanded the levels of use in a manner that is particularly relevant to GIS managers and users. In this research, four levels of use were explored:

1. Use of the system for getting instructions.
2. Use for recording data (e.g., digitizing parcel maps).
3. Use for management and operational control.
4. Use of planning (VanLommel & DeBrabander, 1975).

Finally, Masser and Campbell (1995), in their study of the diffusion and use of GIS in British local governments, noted that "use" is often a problematic measure, depending on the technological and organizational contexts within which a GIS is adopted. It is clear that when "use" is broken down into its various components, it does not make sense for GIS managers to state that their systems are being used. The logical follow-up question to such a statement is asking *how* the system is being used. In other words, is the system being used to its fullest capabilities or is the organization content to use the GIS to perform a few minor functions without really testing its full capacity?

One of the most common (but difficult to assess) measures of IS success is user satisfaction. It is important to distinguish between *system use* and *user satisfaction with the system*. In many cases, a system may be used to a marginal degree without generating much satisfaction by its users, particularly if there are no viable alternatives. In fact, a rule of thumb often suggests that in systems that are underutilized, this state exists because they do not create much satisfaction in their users. User satisfaction, on the other hand, is an extremely relevant measure of implementation success in that it refers to the level of acceptance and positive feelings toward the GIS generated by using the system. It is important to note that user satisfaction has to be assessed after the fact; that is, the users must be in a position to evaluate the data generated, the ease with which they were able to create this data, whether or not the system meets its advertised goals, and so forth. In spite of these examples of issues that can influence user satisfaction, the concept remains very difficult to accurately assess. In effect, user satisfaction asks the manager to get inside the head of the users to see exactly what it is about the system that appeals to them. Even then, a second difficulty relates to determining relative levels of satisfaction—that is, Does satisfaction mean the same thing to different users?

At this point, we need to consider other difficulties with the use of user satisfaction as a measure of implementation success. In particular, two additional questions arise:

1. Whose satisfaction should be measured?
2. How do we separate out individuals' general attitude toward computers from their satisfaction with this specific technology?

In other words, many groups or individuals within the organization could potentially make use of GIS technology. An obvious problem arises when there is a wide discrepancy in terms of their relative levels of satisfaction with the GIS. Why is it working well for some departments and not for others? This question seeks to determine the reasons that the system's introduction is being handled better in some areas of the organization than in others. Another difficulty with measuring satisfaction has to do with the potential for differential satisfaction at various organizational levels. Top management may be extremely pleased with the GIS, while the lower, operational levels are only minimally applying the technology and doing everything possible to avoid using it. Obviously, the "true" measure of user satisfaction in this case rests with the lower level managers and staff who are ostensibly using a system with which they are quite dissatisfied.

The second concern regarding user satisfaction is one that has been well articulated in the IS literature. This problem has to do with the obvious link between user satisfaction and overall attitude (positive or negative) about the use of computers. It is no secret that many individuals in both public and private organizations manifest a great deal of anxiety when confronted with computers. This "computer anxiety" has been the cause of many voluntary withdrawals from companies as departments seek to computerize without spending adequate time retraining employees so that they will know what to expect. On the other hand, newer generations of U.S. and European workers, growing up used to the presence and utility of computers, generally have a much more positive attitude toward their capabilities. This positive attitude often reflects their general level of satisfaction with newly introduced computer systems such as GIS. The point that needs to be stressed for researchers is that prior to studying the implementation of a GIS, it is extremely important that adequate time is spent in assessing the general attitudes about computers on the part of impacted departments. If the perception is that these attitudes are not as positive as they should be, it can have a significant biasing effect on user satisfaction.

Organizations usually do not adopt new, often expensive, technologies simply for their own sake. Rather, as a bottom line, they seek some form of return on their investment. In other words, what is the potential payoff for using the GIS? This point underlines the final two assessments of implementation success: individual impact and organizational impact. "Individual impact" refers to the users' expectation that through use of the new GIS (i.e., through the effect of enhanced information availability) they will derive positive benefits. In other words, using GIS technology in their operations will make them better employees, subordinates, governmental agents, or managers. In determining the individual impact of a GIS there are a number of points that researchers need to bear in mind. For example, "impact" could refer to an improvement in the subordinate's performance in that the GIS allows the employee to make better, more complete, or more accurate decisions. However, impact can also be assessed in other ways. For example, as has been noted,

> "Impact" could be an indication that an information system has given the user a better understanding of the decision context, has improved his or her decision-making productivity, has produced a change in user activity, or has changed the decision-maker's perception of the importance or usefulness of the information system. (DeLone & McLean, 1992: 69)

You can see that, defined in this manner, the concept of user impact is truly multifaceted, comprising not just an assessment that the GIS makes better (more efficient and effective) employees, but that it creates a more professional, well-reasoned workforce, one that is capable of inculcating and using the GIS technology as a stepping-stone to knowledge enhancement. Put in this light, GIS can have the power to teach as well as to nurture better informed workers by allowing these employees to understand and grow with the system and its capabilities. As a result, when assessing individual impact, the manager may be at a disadvantage in knowing at what point to make that determination; that is, when will individual impact be realized to a degree that is noticeable and attributable to the influence of the GIS? We discuss this issue of timing in impact assessment in more detail later in this chapter.

Another aspect of the bottom-line assessment of a GIS's success lies in its relationship to organizational impact. Top management must understandably seek some payoff for their investment of time, money, and human resources in adopting GIS technology. Certainly, within the public sector, it is reasonable to expect municipal authorities to ask some tough questions regarding the expected gains from use of GIS prior to sanctioning its purchase. Campbell's (1993) study of GIS implementation in U.K.

local government demonstrated that one of the most important perceived benefits of the GIS was its ability to improve information-processing facilities. Among the outcomes prized by her respondents was the GIS's ability to improve data integration, speed of data provision, access to information, and increased range of analytical and display facilities.

Unfortunately, the determination of organizational impact through the adoption of GIS technology may be highly uncertain and has been the source of enormous debate within the IS field for years. In fact, many researchers are totally uncomfortable with the idea that an IS must generate some concrete payoff to its organization, preferring to stress other ways in which the system has impacted the organization. For example, they argue, if the system has caused an organization's employees to apply computerized technologies to new problem areas that had previously been ignored or thought inappropriate for its use, the system has demonstrated positive returns. Furthermore, cost-reduction figures may be difficult to generate as they are often measured in terms of time saved through using the GIS over other traditional work methods.

Assessing a system's impact on organizational effectiveness may further be problematic in that it is often hard to separate out the "true" effects of the IS from other biasing or historical moderating effects. That is, the organization is not simply standing still while a new GIS is brought up and running. Other external events are affecting the organization and its ability to function effectively. Consequently, it may be difficult to parse out the "true" impact of a GIS from other activities and events that are influencing the operations of the organization. These and similar issues have combined to make any accurate assessment of organizational impact difficult at best.

The Assessment of Success over Time

One of the points that has been repeatedly stressed in this chapter is the necessity of developing an adequate program in terms of knowing when to determine system implementation success. On the one hand, as previously mentioned, there are definite benefits involved in waiting until after the system has been put in place and is being used by its intended clients before assessing the success and impact of the system. On the other hand, we must be careful not to wait too long to determine system impact and implementation success because other organizational or external environmental factors could influence the organization's operations to the point where we are unable to determine the relative impact of the GIS on operations.

Figure 6.1 illustrates the difficulty faced by the implementation team and researchers (Pinto & Slevin, 1988). This figure shows a simple time line demonstrating the point at which various aspects of implementation success should be evaluated. Note that the time line as drawn has deliberately avoided any specific metrics; that is, we cannot posit the appropriate number of days, weeks, or even months that would necessarily have to elapse for each of these assessments. Rather, the time line simply illustrates the temporal nature of many of the dependent measure assessments. At the earlier stages in the implementation process, the typical assessments of success tend to revolve around issues such as system use and quality; that is, the GIS has been installed and is starting to be used by organizational members who begin to make preliminary evaluations of its quality. (Note from Table 6.1 how each of these issues is more comprehensively defined.) At this early stage, "success" often rests with gaining the acceptance of organizational members to the new system and securing their willingness to actually use it in their activities. Note that it may be too early to make any accurate determination of satisfaction or impact. Rather, at this stage, a number of projections regarding the system's success are being made. The GIS is examined in terms of its technical capabilities—Have pilot project results been satisfactory?—including information use and system quality.

As the time line continues on to the right, additional aspects of system implementation success may be more accurately assessed. For example, once the system is up and running and is in general use, it may now be possible to make some determinations about information "quality." In other words, as we become more knowledgeable about the system due to our continued use of its various features, we are in a better position to accurately gauge the quality of the information that it produces. An

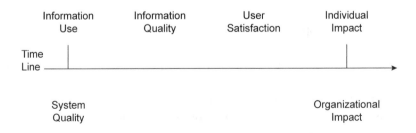

FIGURE 6.1. Assessment of system success over time.

TABLE 6.1. Toward a Unified Model of System Implementation Success

System traits

1. System quality	The system adheres to satisfactory standards in terms of its operational characteristics.
2. Information quality	The material provided by the system is reliable, accurate, timely, user-friendly, concise, and unique.

Characteristics of data usage

3. Use	The material provided by the GIS will be readily employed by our organization in fulfillment of its operations.
4. User satisfaction	Clients making use of the system will be satisfied with the manner in which it influences their jobs, through the nature of the data provided.

Impact assessment

5. Individual impact	Members of the departments using the GIS will be satisfied with how the system helps them perform their jobs through positively impacting both efficiency and effectiveness.
6. Organizational impact	The organization as a whole will perceive positive benefits from the GIS, through making better decisions and/or receiving cost reductions in operations.

acceptably high level of information quality, coupled with the earlier measures of information use and system quality, can be a strong contributor to overall user satisfaction with the system—the state where the user of the GIS has begun to weigh the evidence of system value, and hence is more likely to have positive (or negative) feelings about the GIS. The important point to note here is that research assessing recently installed geographic systems may too quickly seek to acquire measures of user satisfaction before the system's users can reasonably be expected to have enough data or exposure to the system to form a valid opinion.

Finally, at the end of the time line are issues of system impact on individual and organizational operations. As we have noted, "impact" refers to the positive benefits that both individual users and the organization as a whole derive from their use of the GIS. It seems clear that assessments of the benefits derived from a GIS are only apparent following widespread use of the system within an organization. Consequently, as with the case of user satisfaction, we cannot reasonably expect to gain meaningful assessments of individual and organizational impact until enough time has transpired to allow the members of the organization to arrive at informed opinions. Indeed, one should look with a degree of skepticism on research studies that purport to measure impact immediately following installation of a GIS. Such studies are far more likely to be measuring some initial organizational excitement regarding the capabilities of the GIS than they are to measure true impact.

The implications of this model for conducting implementation research are important because they suggest that our goal must be, where possible, to take into account temporal issues when assessing GIS implementation success. While it is beyond the scope of this chapter to posit the precise points in time when each of these various dependent measures should be determined, the time line in Figure 6.1 does serve to illustrate the complexity involved in accurate implementation success measurement. Simply taking measures of these various items at one point in the implementation process, without allowing for the moderating effects of time, may lead to misleading findings and inaccurate conclusions. A more conservative but likely more meaningful strategy is to allow for periodic assessments of various components of system implementation success over a time frame that will allow respondents to more accurately judge such qualitative issues as information quality, user satisfaction, and individual and organizational impact.

Figure 6.1 suggests that while periodic assessments of the current state of the implemented system are important, an accurate determination of the ultimate success or failure of a GIS implementation is equally important. The difficulty lies in attempting to find a suitable reference point when the system has been transferred to the clients, is up and running, and is making some initial impact on organizational effectiveness. One of the important benefits of using such a "postinstallation" system assessment point is that it drives home the point to many implementation researchers and practitioners that the GIS implementation challenge does not end when the system is acquired and set up. In fact, in most instances, there is still a lot of hard work ahead. Implementation managers can foreshorten some of their time with postinstallation user involvement issues depending on the degree to which they and other relevant implementation team members consulted with clients at various earlier stages in the adoption process. However, it is important to point out that an accurate determination of the ultimate success or failure of an IS largely rests with those organizational factors, system use, user satisfaction, individual impact, and organizational impact. The common denominator underlying each of these factors has to do with the desire to fulfill the client's needs: matching the system to the client rather than attempting to alter the client's needs to fit the system.

Implementation Models

GIS are a somewhat unique technological innovation in that they require a combination of both centralized and decentralized processes for their effective implementation through most classes of potential users. The classical implementation conceptual model presumes a centralized structure with a technological innovation originating from some expert source. Under this model, the innovation development process for a community of potential users begins with recognition of a need or problem; moves through research, development, and commercialization of the innovation; continues through implementation and adoption of the innovation by users; and ends with the consequences of the innovation (Rogers, 1983). However, a GIS is a multipurpose tool offering advantages to different classes of users who disperse them at different rates (e.g., utilities vs. planning agencies vs. scientists vs. delivery services). Within each class, considerable adaptation or reinvention appears to occur before the operational characteristics and information product capabilities are perceived as beneficial across the class.

At the adopter level, decentralized implementation processes are often required in order to meet the differing database development needs of each organization, along with the needs of groups and individuals within each organization. Decentralized communication and technology transfer processes occur among and within similar organizations that are all involved to varying degrees in adapting the innovation to their circumstances. While the implementation of GIS hardware and software generally follows the classical model, the implementation of data characteristics and data-handling methods appropriate to the class of potential users (e.g., types of data, quality and accuracies of data, means of collecting data, forms of data storage, forms of system-generated products) probably can be explained best by a decentralized implementation model. Thus, in studying the implementation of geographic information technologies, researchers need to identify both centralized and decentralized implementation processes as well as significant reinvention processes (Rice & Rogers, 1980).

The above discussion suggests that a geographic information innovation, the implementation of which might be promoted or evaluated, could consist of a sole hardware/software combination, a broad range of commercially developed or in-house-developed geographic information-processing capabilities, a unique and useful data set or database method, a standard for data collection, and/or some other technological or institutional development. Thus, an innovation of concern might be any identified innovation that some members of a class of users have found beneficial and that is predicted to be adopted by a high percentage of the social class over time, with or without adaptations to the innovation. In this section, the geographic information innovation typically used for discussion and illustrative purposes consists of computer-based GIS.

GIS Implementation Studies

From an implementation research perspective, it is gratifying to note that a great deal of recent literature in the GIS field has focused on the process of implementation of technological innovations within end-user organizations. Conferences and agenda-setting groups have increasingly expanded the realm of GIS research beyond the original development- and applications-centered descriptive studies to include investigations centered on understanding issues of GIS acceptance and use (Onsrud & Pinto, 1991).

Much of the existing research on implementation of GIS technologies possesses similar properties regarding methodology and research design. Indeed, the vast majority of GIS implementation research consists of single-case studies in which practitioners and academic researchers report on the success or failure of their particular implementation effort within a local government, planning agency, or some other end-user site (Wentworth, 1989). From these case examples, conclusions are drawn that are presumed to be generalizable to the larger population of similar users. For example, Levinsohn's (1989) experience with the introduction of GIS led him to conclude that top management must be involved in major automation decisions. Subsequently, Antenucci, Brown, Croswell, and Kevany (1991) used case illustrations and examples to develop five types of implementation activities: concept, design, development, operation, and audit. As Table 6.2 demonstrates, they were able to make some preliminary distinctions between successful and unsuccessful system introductions on the basis of a host of managerial issues, including planning, staffing, funding, and so forth.

In studying the implementation, use, and assessment of geographic information innovations, there exists a broad range of quantitative and

TABLE 6.2. Elements of GIS Project Success and Failure

	Characteristics of GIS projects	
Activity	Success	Failure
Planning	Rigorous	"Run and gun" style
Requirements	Focused	Diffused
Appraisal of effort	Realistic	Unrealistic
Staffing	Dedicated, motivated, high continuity	High turnover
Funding	Adequate	Inadequate, conjectural
Time estimates	Thoughtful	Rushed or prolonged
Expectations	Balanced	Exaggerated

Note. From Antenucci, Brown, Croswell, and Kevany (1991). Reprinted with permission from Kluwer Academic/Plenum Publishers.

qualitative research methods to choose from. Among these are mathematical modeling, controlled experiments, quasi-experiments, surveys, longitudinal studies, field studies, archival and secondary research, futures research and forecasting, content analysis, case studies, focus groups, and interpretive and critical approaches that have developed in response to shortcomings of the positivist methods (Williams, Rice, & Rogers, 1988). No one method is most appropriate for studying a broad or complex research problem. Each method or combination of methods has advantages and disadvantages as well as different assumptions, biases, and degrees of usefulness (Williams et al., 1988). In fact, a recent approach of research scholars has been to emphasize the use of several research methods in combination in order to balance the weaknesses of each method with the strengths of others (Kaplan & Duchon, 1988).

While current GIS case studies have shed valuable light on some of the important steps in and characteristics of the GIS implementation process, they are hampered by several drawbacks. Case studies and other qualitative forms of social science research have long been criticized for their limitations regarding generalizability to the larger population and their lack of sampling controls (Bonoma, 1985; Cook & Campbell, 1979; Piore, 1979). Furthermore, retrospective reporting of successful or unsuccessful implementation efforts is often subject to considerable information loss and bias, particularly when substantial time has elapsed since the implementation effort occurred. Finally, some of the published case study research involves the reporting of the implementation process by a single individual (usually the manager responsible for the implementation).

Obviously, there is a strong temptation for these individuals to report implementation experiences in the best possible light, even if they are somewhat inaccurate. Wellar (1988b) articulated many of the problems existing in the current research paradigm focusing on qualitative case methodologies.

Some, but not all, of the shortcomings of current GIS case study approaches may be overcome by using more logical and rigorous case research methods. In the MIS literature, Benbasat, Goldstein, and Mead (1987) provided some suggestions on how to conduct and evaluate information system case study research and provided some examples of research areas particularly well suited to case approaches. Lee (1989) and Onsrud, Pinto, and Azad (1992) extended that work to present a scientific methodology with which to conduct case studies of MIS and GIS, respectively. They argued that through an analysis of scientific method—especially the four requirements that a scientific theory must satisfy (i.e., making controlled observations, making controlled deductions, allowing for replicability, and allowing for generalizability)—researchers may identify the point at which scientific rigor is achieved in case study research, beyond which further rigor, particularly at the expense of professional relevance, is questionable. Lee also argued that a qualitative study of a single case can possess more analytic rigor than a statistical study using sophisticated numerical analysis tools. Onsrud and colleagues suggested, however, that the reverse is often more typical. Regardless, important goals for researchers investigating geographic information implementation issues should include becoming familiar with the benefits and drawbacks of the wide range of research methods available, selecting research methods that are appropriate to acquiring the knowledge sought, and selecting a series or suite of methods that are designed to balance the weaknesses of each individual method with the strengths of others.

Although the eventual public and private investment in improved geographic information-handling capabilities is estimated in billions of dollars, few studies have attempted to correlate the usefulness of the technological innovation with adoption, use, and abandonment or to evaluate the efficacy of the technology in social terms. The assumption prevailing among GIS professionals seems to be that because the technology is being adopted, it must be valuable and useful. Yet little information is available, other than in anecdotal form, to support or disprove this assumption. As a result, the study of the implementation of GIS represents an important issue that has not been adequately supported by research efforts to date.

Assessment of Implementation Success

Utilization

One of the more intriguing and continually debated issues in research on
the implementation of innovations concerns the assessment of technological implementation success. In its simplest sense, "success" implies the
degree to which the implementation effort was perceived to be successful.
Such a definition, however, begs the larger questions of determining
exactly what constitutes implementation success and when and how success or failure should be measured. Past implementation studies in other
fields frequently presumed that upon confirmation of the acquisition of a
technological capability, the innovation was successfully implemented.
That is to say, adoption success and implementation success were considered synonymous. This logical error continued even after organizational
theorists argued that the failure of implementation studies to produce
consistent findings was due largely to the failure to identify stages of
implementation in those studies (Greer, 1981; Zaltman & Duncan, 1977).

In evaluating the transfer of geographic information innovations
through a social system, attempts to isolate crucial adoption factors and
processes with their temporal dependencies (i.e., when they occur) and
correlations with organizational attributes (i.e., what organizational factors influence them) will be valuable but will not be wholly adequate. For
reasonable expectations of consistency and generalization of results, it
will be necessary to expand studies to address the full process of adoption
(i.e., acquisition, initial implementation, and use of the innovation by the
organization). Those organizations within a class having already acquired
the geographic information innovation under consideration should be
evaluated in regard to postacquisition implementation, extent of utilization in the organization (e.g., Goodchild & Rizzo, 1987), levels in the
organizational structure of that use, forms of decision making utilizing
the innovation, factors and processes leading to rejections of the previously embraced innovation, and abandonment patterns over time.

One interesting hypothesis long espoused informally among GIS professionals and noted also by technology-transfer researchers in other
fields is that within an organization the influence that various individuals
and groups have on the acquisition of a technological capability may be
only marginally connected to actual staff decisions to use the technology
(Greer, 1981). For instance, the authors can cite an example in which

operational geographic information-processing capabilities were delivered through a contract for an unrelated purpose to the staff of an organization who had little say in the acquisition but who now heavily utilize the system. In yet other examples, the staff appear to have been involved from the very beginning in acquiring the technology, and yet the technology appears to be largely underutilized by staff in the organization. Because GIS implementation is still at the early adopter stage for most classes of potential users, the field is lacking in attempts to study the use of the innovation within and across classes. However, one lesson suggested from past experience in other fields is that after base information is acquired on individual and organizational adopter characteristics, including correlations among the characteristics and with attributes of the innovation, expansion of the field of enquiry to utilization phases may be more fruitful in understanding the implementation process than first trying to directly probe deeper into adoption questions.

Because utility in decision making has proved so difficult to measure, Ives and colleagues (1983) have developed a method for measuring user satisfaction with IS: they argue that user satisfaction may serve as an appropriate surrogate for utility in decision making (see also Igbaria & Nachman, 1990; Raymond, 1987). This approach is also worthy of consideration in the arena.

Impact Assessment

A technological innovation such as GIS is of little consequence until put into general use. Although the effects of technology on both a social system and on the values of that system are extremely important, Rogers (1983) notes that the social consequences of innovations have received very little attention from implementation researchers and change agents. Again, a prevailing assumption has been that if the customs of a social class are altered through general embracement of an innovation, the social consequences, as judged by the adopters, must be beneficial; otherwise, the innovation would not have been embraced. This reasoning in many instances is false, as evidenced in the literature by the numerous examples showing that adoptions of innovations have had highly adverse consequences for an industry or social system as a whole (Niehoff, 1966). The assumption that adoption equals success, in fact, results in a pro-innovation bias. Clearly the findings of social system impact assessments should be held out for consideration by those who have not yet adopted innovations in order to remove the bias and the underlying fallible assumptions. However, it is difficult to measure and evaluate the social and economic responses of classes of users to an innovation. These conceptual and methodological difficulties are at least partially responsible not only for the current lack of research and of generalized findings on social consequences but also for the pro-innovation bias.

Identifying the times at which the effects of innovations should be assessed is difficult. If assessment occurs too early in the implementation process for the class of users, an innovation may not yet have had time to be adapted fully to users' needs. If assessment occurs too late, it may simply memorialize the fact that large numbers of adopters made a wise decision or a poor decision to invest in the particular innovation. Ideally, assessment should occur early enough in the implementation process to offer practical guidance to system designers and later adopters and yet not so early as to result in a gross underestimate of the innovation's effectiveness and value. In yet another sense, innovations such as GIS are continually evolving, and hence there also exists a need to assess them in relation to a social class each time significant technological capabilities relevant to the needs of that class are developed.

One of the obvious problems in attempting to assess the effect of an innovation on a particular organization or across a broad class of users is the difficulty in separating out the effects of the innovation from the effects of other changes in the institution that have occurred contemporaneously with the innovation's implementation and use. Associated with this is the widely acknowledged problem that many of the benefits of IS are indirect and are therefore difficult to measure or estimate quantitatively (Dickinson & Calkins, 1988; Money, Tromp, & Wegner, 1988).

Although a technological innovation may be attractive through traditional cost–benefit analysis, it may have adverse effects on the overall effectiveness of an organization. The reverse is also true. These conclusions suggest a need to consider which individuals or units gain or lose in an organization with respect to the ability of each individual or unit to contribute, the quality of work, and the control of financial and other corporate resources (George & McKeown, 1985; Greer, 1981). Difficulties remain, however, in assessing these factors as well as in determining whether and to what extent such factors are likely to contribute or to distract from the long-term efficacy of an organization or to the long-term efficacy of an entire class of users. Because of this, there is a need to develop alternative methods and strategies for assessing the effectiveness of geographic information innovations at many different levels, such as individual investment in the technology, performance of an entire organization or business, or overall performance of an industry, discipline, or other broad class of users (Bie, 1984; Williams et al., 1988).

Most difficult of all is assessing the societal consequences of an innovation. The societal effects of GISs are potentially very great. These systems have a promising future in helping various segments of society to address some very pressing social problems. Such systems offer efficiencies and capabilities that were previously unavailable and are already being used by individuals, government agencies, private businesses, and a range of organizations to deal with resource management and environmental problems ranging from site-specific problems to global-scale issues. However, even if one were able to confirm that the cumulative effects of decisions made with the aid of GIS have helped increase the overall quality of life and the efficiency of resource production within a social system while decreasing immediate and long-term effects of development on the environment—all of which are far from trivial to assess—adverse and potentially severe social consequences are also likely to arise from implementation of the technology.

One area of concern repeatedly addressed in the geographic information literature are potential shifts in the access rights of citizens to information (Archer & Croswell, 1989; Epstein, 1990; Onsrud, 1989; Roitman, 1988). Many government agencies have established operational land information systems and are making decisions based on analysis of data within those systems that directly affect the daily lives of citizens (e.g., taxing, permitting, service delivery, zoning, districting, and similar decisions). Will such systems increase citizen access to information and promote equal access, or will these systems—particularly in light of recent local government initiatives in the United States to alter state open records laws in the cause of cost recovery or user-fee strategies—create substantial differentials in people's ability to access publicly held information? In addition, the likely effects of geographic information technologies on laws and policies relating to work product protection (e.g., copyrights, patents, trade secrets), rights to privacy, confidentiality, liability, and security have not been widely studied nor have strategies for lessening adverse consequences been fully explored.

Determining what constitutes a beneficial versus a detrimental consequence is a value-laden judgment. Chrisman (1987) suggests that equity is the primary principle around which GIS should be developed, so that everyone affected by use of information in the system will be treated fairly. Rogers (1983) implies that in distributing the consequences of innovations, a strategy should be developed and employed that will decrease, or at least not increase, the magnitude of socioeconomic gaps among members of the social system affected by an implementation program. Geographic information researchers should consider developing goals, policies, and strategies that will promote increased equity in the distribution of the beneficial consequences of geographic information technologies.

Content and Process Models of Implementation

One of the serious problems with past research into the implementation of innovations has been the use of either content or process models as the sole investigative heuristic. A content approach to implementation analysis focuses on determining those specific environmental, organizational, and interpersonal factors that can facilitate or inhibit the implementation process (Leonard-Barton, 1987). Process approaches, on the other hand, strive to analyze the key steps or decisions in understanding how innovations are diffused. While each method is useful, neither offers a complete picture. A thorough approach should identify both the key decision factors in adopting geographic information technologies and the processes by which the implementation occurs.

Content Models

Under a content model approach, data are typically collected from a limited number of case studies in an attempt to identify those implementation model variables that are significant to the adoption process for the particular class of potential users being considered. Past implementation research suggests to us that in the GIS environment we should be particularly cognizant of potential crucial factors in the following areas (Croswell, 1989; Huxhold, 1991; Onsrud, Calkins, & Obermeyer, 1989; Raghavan & Chandf, 1989; Rogers, 1983):

- Visibility of benefits
- Complexity in learning or using the innovation
- "Trialability" of the innovation
- Compatibility with existing values, past experiences, and needs
- Relative advantage of the innovation over the product, process, or idea that it supersedes
- Social norms
- Existence of formal and informal communication channels

- Appropriate balance between mass media and interpersonal communication channels
- The extent and accessibility of vendors of an innovative technology
- Opportunity for information sharing among colleagues
- The presence or absence of champions, reinventors, and opinion leaders in the organization or in the professional community
- How adoption decisions are made within the organization or peer group
- The extent of reinvention necessary to adapt to local circumstances
- The extent of consensus on methods and standards
- Consequences of adopting innovations
- The likelihood of unanticipated repercussions from adoption of an innovation
- Memory of past failures
- The presence of backups if something goes wrong

The above list is not exhaustive. For instance, one potential factor affecting adoption that could be included in the above list is the economic advantage provided by the innovation. In most technology adoption studies to date, however, the economic value of an innovation appears to play a relatively minor role in the decision to actually embrace the technology. Any institution considering investment in an innovation must first cross the threshold of having enough slack in its resources to be able to make some initial investment in it. However, presuming the slack is available, factors other than immediate economic advantage typically are shown to be far more crucial in the actual decision to adopt. Ad hoc observations of adoptions of GIS, for instance, suggest that numerous institutions and organizations are investing in GIS capabilities even though traditional cost–benefit analyses indicate that the investment will never pay for itself over the life of the software–hardware system being purchased. In other instances, although cost–benefit analysis strongly supports an investment in GIS capabilities, the organization has been loathe to incorporate the capability. The differences in the adoption decisions probably can be explained largely by isolating the crucial implementation factors. What at first appears to be an irrational economic decision is converted to a rational decision when the crucial factors in the implementation process are taken into account.

The lists of content factors developed by these and other researchers often vary in degree of comprehensiveness, from broad general outlines (consequences of adopting the innovation) to specific points for consideration (presence of a champion). In spite of their diversity, it is possible to discern some general factors that have been found to be crucial to new system implementation success. These factors were originally posited within the context of new project implementation; however, their general nature makes it possible to draw parallel lessons for the implementation of GIS as well (Schultz, Slevin, & Pinto, 1987). The factors are as follows:

1. *Clearly defined goals* (including the general philosophy or mission of the organization, as well as a commitment to those goals on the part of key organizational members earmarked to use the system). All parties within the organization affected by the implementation of a GIS need to be aware of exactly what tasks they and the new system are expected to perform.

2. *Sufficient resource allocation.* Resources in the form of money, trained personnel, logistics, and so forth are available to support the newly installed system.

3. *Top management support.* Top management within an organization has made its support for the project known to all concerned parties.

4. *Implementation schedules.* A well-detailed plan for new system implementation, including training time, has been prepared and disseminated to all concerned parties.

5. *Competent technical support.* The manager and support personnel for the system installation have the necessary experience and technical competency to ensure a smooth transition to the new system.

6. *Adequate communication channels.* Sufficient information is available on the system's objectives, status, changes, organizational coordination, user's needs, and so forth. Furthermore, formal lines of communication have been established between the implementation team, the system's intended users, and the rest of the organization.

7. *Feedback capabilities.* All parties concerned with the system can review its implementation status and make suggestions and corrections through formal feedback channels or review meetings.

8. *Responsiveness to clients.* Any of the system's ultimate intended users are clients. All potential users of the newly installed system are consulted and are kept up to date on the system's status. Furthermore, they will continue to be assisted after the system has been successfully implemented.

An alternative content model that is "GIS-specific" involves highlighting the critical activities necessary to successfully implement the GIS. For example, a simplified model of GIS implementation could contain the following steps (Tomlinson, 2005):

1. *Consider the strategic purpose.* What are the strategic goals for which we are considering adopting GIS technology? How will we employ it to improve our operations?

2. *Plan for the planning.* Take the time to develop a comprehensive implementation plan to introduce GIS technology. In other words, avoid ad hoc or "evolving" introductions. Rather, address the technology's introduction as a systematic process.

3. *Determine technology requirements.* Do we have the technical means to succeed? If not, what additions to our organization's technology will be necessary to ensure seamless adoption of the GIS?

4. *Determine the end products.* What is the outcome goal we have for adopting GIS? What additional activities do we intend it to support or what deficiencies do we expect the system to correct?

5. *Define the system scope.* This step involves determining what data to acquire, when it will be needed, and the data volumes needed to be handled.

6. *Create a data design.* How will you begin to create the conceptual, logical, and physical structure of the database? This iterative process involves working with the technical people in the organization to determine the physical structure of the data itself.

7. *Choose a data model.* A logical data model describes those parts of the real world that concern your organization. Whether simple or complex, it must mirror the physical world for which it has been developed.

8. *Determine system requirements.* What are the system functions and user interfaces needed to exploit the GIS properly? How can we best link communications, hardware, and software for maximum performance?

9. *Analyze benefits and costs.* Cost and benefits analysis requires us to take into consideration all meaningful elements in the organization's operations regarding the GIS. Thus, we have to understand the cost of operations, data, security issues, staffing, and so forth to generate a clear picture of expected costs and benefits from the system.

10. *Make an implementation plan.* Once all elements in the implementation equation are taken into consideration, it is time to employ project scheduling, scope management, and resource management to create an implementation plan.

As the above list demonstrates, typical implementation efforts often follow relatively similar patterns as far as the most important factors are concerned. Furthermore, as has been mentioned earlier in this chapter, it also becomes quickly apparent that many, if not most, of these factors are more managerial than technical in nature. Implementation theorists and researchers have known for some time that problems with the diffusion and adoption of new technologies are often based on human issues rather than on technical difficulties or concerns. This is not to suggest that a system does not need to be technically adequate in order to be accepted. However, as Schultz and Slevin's (1975) development of the ideas of organizational validity and acceptance demonstrates, the battles for successful information system implementation are usually won or lost, not in resolving all technical issues relative to the GIS, but in appealing to and attempting to address organizational members' concerns.

As already mentioned, the problem with the content model approach is that—while it offers important information for managing a system's implementation—it is essentially a static representation of the implementation effort. In other words, content models do not mirror the importance of the process by which a new system is implemented. Most definitions of implementation have included as part of their description phrases such as "the process of organizational change" or a "change process." Consequently, implementation models need also to reflect the dynamic nature of new system diffusion. In other words, the content model approach has value and provides insights but should be used as only one of the components in a comprehensive implementation model.

Process Models

Unlike content models of implementation, which are aimed at identifying the factors that are the key determinants of innovation acceptance and use, process models are concerned with determining the key phases in the adoption process. One model suggests that there are two main subphases in the innovation process:

1. *Initiation.* The organization becomes aware of the innovation and decides to adopt it.
2. *Implementation.* An organization engages in the activities necessary to put the innovation into practice and incorporate it into existing and developing operations.

These terms, "initiation" and "implementation," have been redefined by Schultz and colleagues (1987) as the "strategy" and "tactics" of an implementation effort. Their argument suggests that one simple but effective way to view the implementation process is as a distinction between planning activities and action-oriented efforts. That is to say, planning activities (termed "strategy") are related to the early planning phase of the implementation process. They represent either the conceptualization of the new system implementation or of its planning and control. A second set of factors (referred to as "tactics") is concerned with the actual process, or the action of the implementation, rather than its planning.

The essence of Schultz argument is that conceptualizing system implementation as a two-stage process has further implications for system performance. Figure 6.2 shows the breakdown of strategy and tactics by low or high score depending on the level to which these issues were addressed in the implementation process. For example, a high score on strategy would imply that the strategy was well developed and effective. This value could either be assessed in a subjective (or in an intuitive) manner or more objectively using some surrogate measures of the implemen-

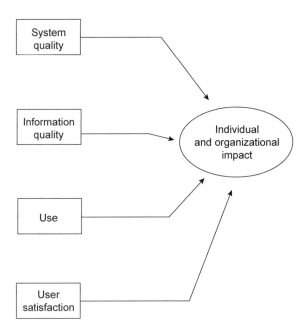

FIGURE 6.2. Modeling the determinants of system implementation success.

tation process (such as initial organizational member buy-in, schedule and budget adherence, etc.). A manager could, for example, determine that the strategy was deficient based on past experience that he or she had with other, successfully implemented systems.

We can speculate on the likely outcomes for new system implementation efforts, given the assessment of their strategic and tactical performance. Figure 6.3 illustrates the four possible combinations of evaluated performance of strategy and tactics. The terms "high" and "low" imply strategic and tactical quality—that is, the effectiveness of operations performed under the two clusters.

Four types of errors may occur in the implementation process. The first two error types were originally developed in the context of the statistical testing of hypotheses. The last two error types have been suggested as the result of research on implementation. Type I error occurs when an action should have been taken but was not. Type II error is the taking of an action when none should have been taken. Type III error means taking the wrong action (solving the wrong problem). Type IV error occurs when a solution is discovered but is not used by an organization (Schultz & Slevin, 1975). Each of these error types is more or less likely to occur depending on the mix of strategy effectiveness and tactics effectiveness. Knowing the interaction of strategy and tactics and their probable effect on project success and potential error type is important in understanding the implementation process.

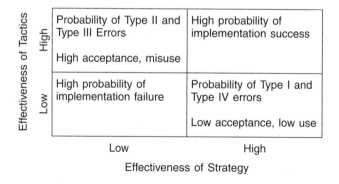

FIGURE 6.3. Strategy/tactics effectiveness matrix. From Schultz, Slevin, and Pinto (1987). Copyright 1987 by Randy L. Schultz, Dennis P. Slevin, and Jeffrey K. Pinto. Reprinted by permission.

High Strategy, High Tactics

Quadrant 1 in Figure 6.3 holds those systems that have been effective in carrying out both strategy and tactics during the implementation process. Not surprisingly, the majority of implementation efforts under this classification are successful.

Low Strategy, Low Tactics

The reciprocal of the first quadrant is in the third, where both strategic and tactical functions were inadequately performed. We would expect implementations falling in this quadrant to have a high likelihood of failure.

Low Strategy, High Tactics

While the probable results of new system implementation are intuitively obvious in quadrants 1 and 3, the results for efforts that fall in quadrants 2 and 4 are not. Quadrant 2 represents a situation in which initial strategy functions were insufficient, but subsequent tactical activities were highly effective. Some of the expected consequences for system implementations falling into this category would be an increased likelihood of Type II and Type III errors. Type II error would occur in a situation in which the initial strategy was ineffective, inaccurate, or poorly developed. However, in spite of initial planning inadequacies, goals and schedules were operationalized during the tactical stage of the implementation. The results of Type II error could include implementing a poorly conceived or unnecessary system that has received no initial buy-in from potential users and either may not be needed or will not be used.

Type III error may also be a consequence of low strategy effectiveness and high tactical quality. Type III error has been defined as solving the wrong problem, or taking the wrong action. In this scenario, a need has been identified or a new system is desired, but owing to a badly performed strategic sequence, the wrong problem was isolated and the subsequently implemented GIS has little value in that it does not address the right target. In this case, the tactics to develop and implement a new system are again well conceived, although initial planning and problem recognition were poorly done. The result would be high acceptance of the system, as well as its general misuse, owing to action being taken to implement a system when none may have been warranted.

High Strategy, Low Tactics

In quadrant 4 lie cases where strategy was effectively developed but subsequent tactics are rated low or ineffective. One would expect implementation efforts classified in this quadrant to show a likelihood of Type I and Type IV errors (i.e., not taking an action when it has been determined that action is needed or simply not using the new system). To illustrate, consider a situation in which strategic actions have been well performed, thus suggesting the installation of a GIS. Type I error would occur when little action is subsequently taken and the tactical activities are so inadequate that the new system is not implemented.

Finally, Type IV error would occur following an effective strategy that has correctly identified the need for a GIS for an organization, but, after poor tactical operationalization, the system was not used by the clients within the organization for whom it was intended. In other words, Type IV error is the result of low client acceptance. As discussed earlier, the reasons for lack of acceptance are numerous, but most often they revolve around a failure to match the system to the needs of the organization and its specific personnel.

Process models of innovation have been useful for identifying the important steps in gaining acceptance and use of new innovations. However, process models, by themselves, are also afflicted with some difficulties (Srinivasan & Davis, 1987). First, as mentioned earlier, most process approaches in the past have been highly qualitative in nature. Second, they do not attempt to determine who the key players within an organization are at each step in the adoption decision. Third, these models may vary or fluctuate to a significant degree depending on the type of innovation that the organization is trying to adopt—in other words, the models may not be sufficiently generalizable. Fourth, it is often the case that the desire for innovation may exist within an organization at a particular node, whereas the overall organization is not particularly innovative. For example, certain nodes of an organization, such as engineering or software development, may seek to adopt a GIS, while the overall organization is reluctant to pursue computer information system innovations.

Implications for Implementation Research and Practice

The discussion in this chapter has highlighted a number of salient issues that GIS researchers need to consider when attempting any sort of geographic system implementation assessment. These points are enumerated below.

1. *Installation is NOT implementation.* One of the enduring difficulties with much of the research that is ostensibly done on GIS implementation is that it is essentially mislabeled: it could more properly be termed "installation" research. The problem is that with many researchers, there is a basic difficulty in distinguishing between the activities necessary to acquire and install a GIS and those that pertain to its implementation within the organization. A paper on GIS diffusion by Onsrud and Pinto (1993) attempted to make this point by distinguishing between a content model of GIS implementation (i.e., those organizational and behavioral factors that contribute to implementation success) and a process model of the acquisition and installation process. A first, necessary condition of implementation research demands that a study's principal authors first make clear in their own minds the distinctions between those factors important to acquiring a GIS and those that relate to its implementation.

2. *Any IS is only as good as it is used.* What is the bottom line for any implementation effort? If we are willing to accept that "successful" implementation can be measured on the basis of installing and bringing a system online, without any subsequent regard for its acceptance and use, we are misinformed. As Schultz and Slevin (1975) noted, the technical validity of a system is only the first step in the process of implementation. It is in successfully navigating the idea of organizational validity that we begin to understand the role of the client in determining success.

One of the most important points that past research and experience have taught us is that the client is the ultimate determinant of successful system implementation. This lesson, so fundamental to practicing managers, is one that continually escapes the attention of researchers and IS theoreticians and must be continually relearned. We continue to deal with a culture that is fascinated with the latest technological advances. One has only to pick up a newspaper or business magazine to see wide reference being made to the impending "information superhighway," despite the fact that the average person may be over a decade away from realizing many of the practical benefits of this family of technologies. In our drive to innovate, there is a very real danger that we will begin to pursue technology for technology's sake, rather than working to create systems that have practical and useful features. The oft-repeated statement "Of course it will be used—it's state of the art!" reveals a high level of naiveté about how innovations are received and perceived by the average person. It is vital that GIS implementation researchers understand the importance of client acceptance and use as a necessary condition of implementation success.

3. *Successful implementation may require system and client modification.* A fundamental finding of Schultz and Slevin's (1975) earlier work referred to the concept of "organizational validity." What is important about this concept is the implication that any new GIS must be "right" for the organization toward which it is targeted. Furthermore, the "right" system refers to the importance of matching the needs and attitudes of the client organization to the new technology. Many firms acquire GIS that are underutilized because they were inappropriate for the target organization. However, it is important that the reader understands that an "inappropriate" system is often not the direct result of technical difficulties or performance characteristic flaws. Many times a GIS will be perceived as inappropriate because it does not conform to the attitudes and value systems of the majority of organizational members. In other words, the cultural ambience of a firm must be taken into consideration when determining its technological needs.

We argue that the process of developing greater organizational validity for a GIS innovation involves a process of mutual adaptation between the system and the client organization. Significant preliminary work is required from the project manager and team members as they scan the client and objectively assess attitudes and needs regarding an IS. If the team determines that it is not feasible to implement a GIS within the current organizational context, they need to begin formulating plans for how to create a more supportive environment. That process may require either modifying the GIS to suit the technological needs of the company, engaging in large-scale training programs within the company to create an atmosphere of acceptance, or both. Unless project managers and their teams work to address potential problems of organizational validity, their highly developed and technically sophisticated systems are likely to fail without having been given a sufficient chance to succeed.

4. *When assessing "success," give the system time to be incorporated into the client's operations.* In assessing the performance of an IS implementation, GIS researchers are faced with a significantly complex task. Because so much depends on the acceptance and use of the system and its resultant impact on the client organization's operations, it is often of very questionable utility to assess the system's impact before it can, in some sense, be determined. Too often we have sought quick assessments of system impact as part of a general data-gathering process, whether "impact" was yet a

meaningful variable or not. A more appropriate methodology might be to sample an organization's implementation effort at various points over some extended time frame following installation to gain better informed data from the sample population being researched. Put another way, it may be possible to measure, with reasonable accuracy, such issues as system performance and client use at an early point following installation because they are more immediately derived success measures. On the other hand, any estimates of system impact made too early in the implementation process are likely to be misleading at best and utterly wrong at worst.

The obvious difficulty with making any sort of post hoc impact assessment is that of determining when best to make such an analysis. In other words, How long should a system be in place before it is analyzed for its utility to organizational operations? Any new system will require a "shakedown" period while clients learn how to use it and adapt it to their activities. At the same time, they are also likely to be learning the various strong and weak points of the GIS and hence are usually unable to formulate an accurate assessment of the system for some time. On the other hand, the longer they take to assess implementation and performance success, the greater the likelihood that other intervening variables will interfere with their ability to give an honest appraisal of the GIS. For example, as new technological breakthroughs occur, any current GIS will begin to look old and increasingly cumbersome to its users, particularly when compared with the capabilities that new systems offer.

Clearly, a trade-off must be made between assessing implementation success too quickly and waiting too long. Whatever decision rule is adopted by GIS researchers, they need to make it with due regard to the various trade-offs that exist and the decision criteria must be applied consistently across all implementation efforts of a similar nature.

5. *Consider who stands to gain and lose from assessing GIS "success"–remember the politics of the organization.* An important complicating factor needs to be considered when seeking to determine GIS implementation success: the problem with attempts to develop a rational approach to assessing IS implementation success in the face of the irrationality that often accompanies organizations. This irrationality is usually manifested in examples such as the power and politics that accompany organizational activities, where one or more parties are intent upon furthering their own goals, even at the expense of the overall organizational good (Pinto & Azad, 1994). Normative models of how organizations ought to function are often notorious for failing to describe reality. Consequently, one facet

of determining successful GIS implementation must be to consider the potentially self-serving effects of any party's willingness to label a GIS as either successful or failed. For example, from a power position, is it expedient for one organization actor to dismiss a newly installed GIS because he or she perceives that it furthers the goals or power of another departmental manager? Within the context of geographic information sharing, Masser and Campbell (1995: 247) made a similar observation in noting that "organizational and political factors apparently offset in many instances the theoretical benefits to be obtained from structures that seek to promote information sharing." There is clearly no reason to suppose that those same factors will not affect attempts at posing overly rational methods for obtaining information on new system implementation success.

Conclusions

For managers attempting to better understand the management of GIS within their organizations, a basic knowledge of organization theory and human behavior is essential. In this chapter we have argued that many, if not most, of the problems associated with managing the introduction and use of a new IS are people problems rather than problems associated with technical difficulties. Implementation theory and research have for years known that the most prevalent implementation problems (such as lack of acceptance and use) are the result of poor development of an organization's human assets. Consequently, any discussion of the process by which a GIS is introduced and managed within an agency or organization must be predicated on developing a greater understanding of the organization as a social system.

Chapter 7

Organizational Politics
and GIS Implementation

There is still far more to accomplish within the technical realms of GIS: algorithms, user interfaces, temporal databases, efficient storage schemes, better raster–vector integration, and so on. Nevertheless, GIS technology continues to be acquired and put into use by a wide variety of organizations. Unfortunately, the technology implementation process in the organizational context remains riddled with problems leading to project slowdowns or outright discontinuance in many instances. This trend is usually attributed to a wide range of behavioral and organizational difficulties that tend to impede the more effective use of GIS in organizations.

This chapter focuses on one of the more profound and, in many ways, fascinating (and understudied) themes in successfully implementing new technologies such as GIS: the role of organizational politics. Although anecdotal evidence and case histories abound that link politics to both GIS implementation success and failure, we lack a thorough understanding of the impact of corporate politics in the GIS implementation process. In effect, no organizing framework has been advanced to suggest the ways in which politics can help or hinder the implementation of GIS.

As a first step toward building such a framework, our mission in this chapter is to advance the "positive" management of organizational political behavior (OPB) as an integral part of GIS technology implementation process. Toward that end, this chapter attempts to address the following five goals: (1) review the evidence on OPB and information and GIS technology implementation; (2) provide logical propositions as to why OPB takes place; (3) establish the contents of OPB in analytical terms (i.e., When can

Portions of this chapter are from Pinto and Azad (1994). Copyright 1994 by the Urban and Regional Information Systems Association. Adapted by permission.

114

an organizational behavior be considered political?); (4) put forward a normative view of "positive" OPB in the form of a number of managerial actions to promote the likelihood of successful GIS implementation; and (5) present two GIS implementation minicases from state agencies that illustrate and refine the analytical and normative aspects of the OPB framework provided.

Organizational politics and political behavior has been defined in a number of ways, but these definitions often have some underlying common themes. For our purposes here, we define *organizational politics* as any process by which individuals and groups seek, acquire, and maintain power (Pinto, 1996). By natural implication, knowledge and information, such as that contained within a firm's GIS, represent a significant source of power and hence are prone to provoke a variety of political and power behaviors. As we will see, these behaviors may be the result of attempts to "corner the market" on GIS access within an organization, to limit its use by certain people, and so forth. All such political behaviors have an enormous impact on the ability to acquire, successfully implement, and operate a GIS. "Managing GIS," therefore, often consists of first successfully identifying various political actors and forces and then managing and minimizing their negative effects within an organization's political arena.

There is a widespread belief that technical issues and decisions are at odds with political behavior. However, this appears to be too simplistic an argument to characterize reality. Furthermore, the real issue appears to be not that the "technical" and the "political" do not mix—often they do by reactive default rather than through proactive management—but that in organizations the "political" process has to be "positively" managed much like any other organizational concern. Let us look briefly at three well-known cases that help us better understand the core problem.

The *Challenger* Disaster

On the morning of January 28, 1986, schoolchildren around the country watched as, after repeated delays, the space shuttle *Challenger* finally lifted off. Seventy-three seconds later, *Challenger* disappeared in a raging fireball. The Rogers Commission, appointed to investigate the disaster, determined that the immediate cause of the explosion was physical: two O-rings designed to seal joints on *Challenger*'s right booster rocket had failed.

The more fundamental cause of the accident, however, was *organizational and political* (Hult & Walcott, 1990). The Rogers Commission called the decision-making system for the shuttle program "clearly flawed." The decisions by NASA management and the contractor were influenced by myriad (political) factors that combined to produce the fateful *Challenger* incident: "turf" consciousness among the three space centers, inadequate communication of the technical uncertainty associated with O-ring risk, and congressional/public pressure to produce results, to name just a few.

Indeed, the *Challenger* episode is a dramatic illustration of the issues that concern us in this chapter and is of importance to an accurate understanding of the political behavior in organizations during the implementation of complex technical projects.

Xerox Alto

If we ask consumers and users what names they associate with the multibillion-dollar personal-computer market, they will answer Dell, Gateway, Apple, or Hewlett-Packard. But no one will say Xerox. Twenty years after it invented personal computing, Xerox still means "copy" (and printers).

But in 1973, many years before Apple, IBM, or Tandy released their first personal computers, scientists at Xerox's Palo Alto Research Center (PARC) produced the Alto, the first computer designed for personal use. By 1976, still well in advance of any other enterprise, PARC's brilliant team had completed a system of personal computing hardware and software. It was not matched in the marketplace until 8 years later with the appearance of Apple's Macintosh computer—a product whose intellectual roots, ironically, belonged to PARC and therefore to Xerox.

Yet those at PARC who expected Xerox to capitalize on their extraordinary inventions remained frustrated throughout the 1970s and 1980s by the other workgroups at Xerox who backed a far more elaborate office computer system and then failed to introduce it until Apple and IBM had already set the standards of the marketplace. In *Fumbling the Future: How Xerox Invented, Then Ignored, the First Personal Computer*, Smith and Alexander (1988) tell a compelling tale of how innovation within large corporate structures can be miscalculated and mishandled when organizational politics and culture are not positively dealt with by management and professionals.

Airbus A-380

Conceived at the turn of the millennium to be the largest airplane ever developed, the A-380 program has been one that is characterized by technological snags and political missteps nearly from the beginning. Designed to seat 555 people in a two-deck configuration, Airbus developed the project as a new venture to provide long-haul airlines a means to cross large distances while carrying hundreds of passengers. Airbus first began studies on a very large 500-seat airliner in the early 1990s. The European manufacturer recognized that by developing a competitor to the successful Boeing 747 they would be able to end Boeing's dominance of the very large airliner market and round out Airbus's product lineup.

Airbus began engineering development work on such an aircraft, then designated the A3XX, in June 1994. The aircraft was originally intended to enter commercial service in March 2006 with a final price tag of $16 billion. Unfortunately for Airbus, technical problems ballooned as the aircraft moved closer to commercial launch. Current estimates place the A-380 at least 2 years behind schedule and approximately $4 billion dollars over budget. Among the chief reasons for the delays and technical problems are the political and social differences among the various European firms supplying parts for the aircraft. Most recently, companies in Spain, France, and Germany discovered that as a result of using incompatible design software, miles of cable and wiring are having to be restrung (in some cases, by hand) on aircraft hung up in production. The political consequences of these errors and coordination problems led first to the resignation of the A-380 program head and most recently to the ouster of Airbus CEO Christian Strieff. Strieff admits that he underestimated the backlash to his agenda to make the aircraft manufacturer more efficient by closing plants, reassigning work, and renegotiating labor contracts. In effect, organizational politics resulted in huge delays in a major program for Airbus at a time when it could least afford them (Matlack, 2006).

Why Organizational Politics Matter

What do these "technical" failures have in common? One common theme is the lack of attention to political issues in organizations and their "positive" management. From where does this lack of attention originate? According to Norton Long (in Pfeffer, 1992: 29), "People will readily admit that governments are organizations. The converse—that organizations are governments—is equally true but rarely considered." While it rings true of everyday reality, it is not a given for many Americans, who generally are ambivalent about power and politics or with its dynamics. At one level, there is often a distrust of the motives of workplace people who actively seek power and thrive on politics. At another level, there is a recognition that political behavior is normal. In plain terms, the attitude and feeling toward power (perhaps this can be traced back to the nation's birth) is negative but tolerated.

A study of organizational politics (Gandz & Murray, 1980) surveyed 428 managers. Their answers illustrate the ambivalence concerning politics: 90% of the respondents said the experience of workplace politics is common, 89% said that successful executives must be good politicians, and 76% said that the higher that one progresses in the organization, the more political operations become. However, 55% of the same respondents said that politics were detrimental to efficiency. These figures leave an impression that we know politics exist, we grudgingly admit that they are necessary, but nevertheless we do not like them.

There are other reasons for this perspective toward politics. Pfeffer (1981: 12) has eloquently offered the following reasoning:

> To socialize students in a view of business that emphasizes power and politics would not only make the compliance to organizational authority and the acceptance of decision outcomes and procedures problematic, but also it might cause recruitment problems into the profession. It is certainly much more noble to think of oneself as developing skills toward the more efficient allocation and use of resources—implicitly for the greater good of society—than to think of oneself engaged with other organizational participants in a political struggle over *values, preferences,* and *definitions of technology.* (emphasis added)

Regardless of the reasons for this disposition toward politics and power in organizations, one of its impacts is that as a topic for study and dialogue, beyond the field of political science, it has not received much attention, even in the managerial circles of organization behavior and organization theory (notable exceptions are Bacharach & Lawler, 1980; Ferris & Kacmat, 1992; Pettigrew, 1973; Pfeffer, 1981, 1992; Mintzberg, 1983; Yates, 1985). This lack of attention to the topic is not limited to management. Benveniste (1989), Forester (1989), and Fischer and Forester (1993) have offered critiques of this trend and attempted to reverse it in the urban planning arena. The fact that the subject of power and politics in organizations has received less attention than it deserves merely adds to the confusion and misunderstanding surrounding the topic. In particular, we view this misunderstanding as a liability for GIS managers, planners, and professionals engaged in technology implementation, given the "technical/rational" nature of discourse on the GIS-related matters.

The steps toward development and use of GIS in many organizations can be conflict-laden. Those who push the theme that GIS fosters cross-functional cooperation and integration by default gloss over deep social and value conflicts that social change due to GIS implementation may precipitate. In practice, organizational participants can have major "battles" about what kind of hardware platform to acquire, how to organize GIS data layers, degrees of spatial database accuracy, the means of distribution of processing power (workstation vs. centralized), and the standards to govern data exchange (personal communication, W. Huxhold, 1993).

Two case studies at the end of this chapter illustrate the typical conflicts among functional workgroups within state departments of transportation—choice of GIS software is one example. In one instance, the planning group structuring their work around one particular software product were at odds with the engineers and construction group that had a history of using a separate product. However, the data-sharing requirements and the assumed cost savings had forced the issue of a standard software platform across the agency on every workgroup's agenda. One of the biggest issues of concern was the conversion of the so-called macros in the different languages that staff in both workgroups depended on. Each viewed the costs of conversion as an unnecessary burden.

What is important to understand is that these situations of conflict cannot be written off as flukes or exceptions. In fact, a phenomenal amount of anecdotal and case history information exists that strongly reinforces the importance of understanding and *effectively utilizing* organizational politics as a tool in successful implementation. For example, Croswell (1991) documented some of these in his survey of GIS and related publications. Despite this "stylized fact," the GIS research community has been slow to disentangle the web of organizational politics and translate it into usable positive managerial actions that increase the probability of implementation success. However, GIS is not alone in this regard. The study of organizational politics is only marginally better understood in similar situations, such as introduction of innovative IT in organizations (Danziger, Dutton, Kling, & Kraemer, 1982; Frantz & Robey,

1984; Kraemer, Dutton, & Northrop, 1981; Markus, 1983; Mumford & Pettigrew, 1975; Pettigrew, 1973; Robey, 1984; Robey & Markus, 1984).

The rest of the chapter is devoted to some major questions as an attempt to uncover the dynamics of OPB and GIS implementation:

- Is there any evidence to show the impact of OPB on information and GIS technology implementation?
- Why does OPB take place?
- What are the analytical contents of OPB?
- What are some concrete OPB steps that can enhance the success probability of the information and GIS technology implementation process?
- Can we illustrate the impact of OPB on the GIS implementation process through concrete evidence?
- What are the major conclusions and directions for further research?

We address these questions in sequence in following sections.

Politics and Public-Sector Information Technology

We are interested in finding out how "positive" OPB can contribute to the success of GIS implementation. From a research point of view, this can be translated as follows:

1. Project success is the dependent variable.
2. OPB is one of the independent variables during GIS implementation.
3. The dynamics of OPB and implementation process interaction is also a determinant of success.

Although the issues in the latter category are interesting, they fall beyond the scope of this chapter. We are more concerned with the second group of issues, or more concisely, what are the contents of OPB?

Perhaps due to the present early stages of research on GIS implementation, researchers in studying GIS implementation have not addressed the second question, or at least not in enough depth to illuminate the complexities of the topic. For example, Budic (1993) confirmed earlier assertions about the importance of political backing in GIS acquisition

(Godschalk, Bollen, Hekman, & Miles, 1985; Croswell, 1991; Sommers, 1990), asserting that political support was rated very highly for incorporating GIS technology within government agencies among four states in the southeastern United States. Similarly, Campbell (1991) found that political factors were important in GIS project implementations in two U.K. local authorities.

However, as mentioned earlier, GIS is not alone in this respect. The research on IT implementation has paid little more attention to OPB. No matter how scant this research is, it is useful to review its major themes in the hope of gaining insight into OPB during GIS implementation. In general, most research in this area has been concerned with uncovering some form of political (intangible upper-management) support in successful IT implementation (see the inventory of survey research by Kraemer & Dutton, 1991). A very small portion of the research has concentrated on political organizational behavior and IT implementation (Frantz & Robey, 1984; Kling, 1980; Markus & Bjorn-Andersen, 1987; Robey, 1984; Robey & Markus, 1984).

First, we consider the latter body of research. We offer an interpretation of the evidence based on two streams of research: (1) the "political impact" school of IT in organizations and (2) the "conflict resolution" model of IT in organizations. Although our approach of applying "positive" OPB in the IT/GIS implementation process for higher success rate is quite distinct from these streams of research, we are closer to the latter than to the former.

Political Impacts of IT and GIS

Political Impact School

A small but thriving research track on politics and IT implementation in the United States has been the so-called impact school. In other words, there are important considerations of how the implementation of IT changes social and power relations in the organization. Researchers at the University of California at Irvine (originally the URBIS group, now known as CRITO) are the main adherents to this view and have produced evidence to support their assertions (Danziger et al., 1982; Kraemer et al., 1981). In a nutshell, Kraemer and his associates have opposed the characterization of IT as an instrument by which different organizational goals might be accomplished. In their view, this image can lead to the incorrect conclusion that IT is simply a neutral tool in organizational life to be used as best fits the organization. In fact, they postulate that the potency of IT in decision making, if not in other areas, makes it politically important. According to Kraemer and associates, the political significance of IT arises from three features of IT use.

First, they argue that there is the political significance of the outputs produced by IT. Information per se is not power, but those with the "best" information are often successful at accomplishing their objectives. Depending on how IT and information delivery is organized and provided, different individuals and factions in organizations can gain or lose power relative to others. This is especially true in the context of the contributions of IT to decision making mentioned above.

Second, there is the "resource politics" of IT, arising from the fact that those who control IT govern a large share of organizational resources. Control over these resources brings power, both through building a base for further increases in demands on resources and through control over capabilities that others in the organization or its clients need.

Finally, IT brings "affective power" with its inherent attractiveness as an activity. Those who are engaged in IT are perceived by many as advanced, sophisticated, and professional. Also, since many people are intimidated by technical jargon and IT outputs, these can be used effectively to obfuscate the underlying issues in disputes and to weaken opposition.

Overall, Kraemer and associates have focused on the aggregate organizational impacts of IT. This is because, in their view, the fundamental question about IT and organizational politics is who gains and who loses from IT. To simplify, the "dependent variable" in their framework is "redistribution of power." The input (independent) variable appears to be the implementation of IT itself. Then these inputs to the process are moderated to redistribute power by (1) organization of production/delivery for information in the organization; (2) control over IT-related resources; and (3) symbolic value of being associated with high-profile IT-related activities.

Figure 7.1 is a "distillation" of their framework and major conclusions. It is clear from our characterization that their concern is that IT implementation leads to political shifts and distribution of power—therefore, our label "political impact." They are not concerned (as directly as us) with the use of positive OPB to enhance the chances of successful IT implementation. An important aspect of their research—to most GIS professionals—is that it is conducted solely based on public-sector data and evidence since its inception in the early 1970s.

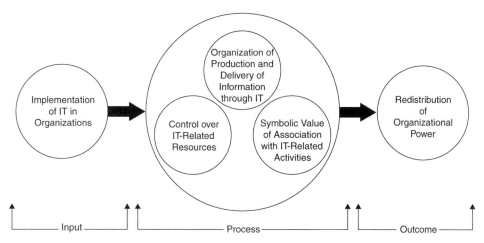

FIGURE 7.1. Essence of the Kraemer and associates' framework: "The political impact" of IT implementation. From Pinto and Azad (1994). Copyright 1994 by the Urban and Regional Information Systems Association. Reprinted by permission.

A subsidiary point to our concerns, but an interesting one in their conclusions, is the so-called reinforcement politics. That is, they provide counterevidence to the two prevailing views on IT impact: some have predicted that IT will alter the political profile of organizations by shifting power to technocrats—this view emanates largely from the literature on the role of experts in organizations (Crozier, 1964; Downs, 1967b; Ellul, 1964); and others have suggested that IT can strengthen pluralistic features of organizations by providing interest groups with the ability to respond to their opposition with the tools of technology—this view is largely influenced by the political science models of organizational behavior (Allison, 1971), interest group competition, and public choice models of public administration. However, the UC-Irvine group maintains that IT has reinforced the status quo by providing the existing power elite with the tools to perpetuate and strengthen their power. The evidence presented by them suggests that this has been the most common outcome of IT.

User Participation (or Conflict Resolution) School

Another stream of research also explicitly incorporates politics into its framework and results. The major focus is on user participation in IT implementation, casting it as a political process. The core element of this approach is its focus on differences among the expectations and interests of stakeholders, attributing the general dissatisfaction with IT implementation to unmet stakeholder expectations (Markus, 1983). This literature has provided a starting point for understanding the politics of IT implementation (Frantz & Robey, 1984; Kling, 1980; Markus & Bjorn-Andersen, 1987; Robey, 1984; Robey & Markus, 1984) focusing on the strategies and tactics used by stakeholders to influence the IT implementation process in their favor. Because the stakes in IT implementation are usually high and have long-term consequences—the "moderator" variables according to Kraemer and associates—a high level of political activity during IT implementation can be expected (also identical to Kraemer et al.'s conclusions).

However, this stream of research itself can be divided at least into two groups: "zero sum" and "non-zero sum" categories. There are those who treat IT implementation as a purely political process with clear disregard for legitimate organizational goals. This position tends to mistrust all appeals to organizational goals and to suspect that individuals are motivated *only* by their own interests. The complicating factor in IT implementation is that, because advance demonstration of universal benefits is always problematic, the legitimacy of organizational goals is hard to establish objectively (Kling, 1980; Mowshowitz, 1981). Thus, according to this group of researchers, conflicts during IT implementation may be viewed as "zero-sum" games in which the gains won by one party must equal the losses suffered by another.

The other group of researchers is less skeptical of the political model and espouse a "constructive" *conflict resolution* mode of IT implementation. They conceive of IT implementation as a "non-zero-sum" game wherein multiple parties can come away satisfied. This line of inquiry grounds itself in the management literature on resolution of conflict as an essential skill and process characteristics (Filley, 1975; Mintzberg, 1983; Robbins, 1992; Slevin, 1989), thereby making management of conflict during IT implementation a central piece of the puzzle for more successful systems. Thus researchers put forth the positions that, despite the presence of conflicting interests among the stakeholders in IT implementation, it is conceivable that managers could facilitate the resolution of conflicts and produce a "win-win" outcome, deemed successful by all parties.

Robey and associates have tested several versions of this conflict resolution model (Robey, 1984; Robey & Farrow, 1982; Robey, Farrow, &

Frantz, 1989; Robey, Smith, & Vijayasarathy, 1993). The model consists of four variables: participation, influence, conflict, and conflict resolution. Participation is treated as a determinant of influence, and influence is treated as a determinant of both conflict and conflict resolution. Overall, the results of the model support the key role of participant influence and conflict during IT implementation. Figure 7.2 illustrates the structure of this model.

According to Robey and colleagues (1993), given the "realistic" assumption that stakeholders will disagree on fundamental issues during an IT implementation project, it is important to understand the manner in which conflicts are managed. One approach smoothes over conflicts by minimizing disagreements among participants. This can reduce conflict in the short term, but in the long run it may result in important issues going unaddressed. If conflicts are encouraged to surface and then be resolved constructively, project success is likely to be greater.

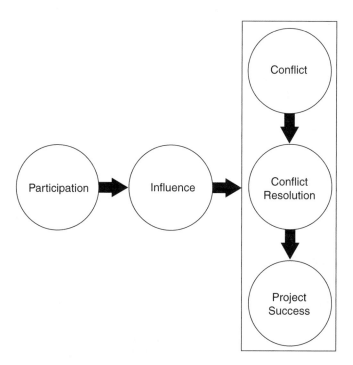

FIGURE 7.2. A sketch of the Robey and associates' conflict model of IT implementation. From Pinto and Azad (1994). Copyright 1994 by the Urban and Regional Information Systems Association. Reprinted by permission.

Robey and his associates' results provide convincing evidence of the presence of conflict during implementation of IT, and the general means to its resolution through influence by users participating in the process. However, the model has limitations in two important respects. First, as we will see later, the presence of conflict and influence are only two manifestations of OPB. Second, the resolution of conflict in the organizational context may be an excellent explanatory variable, but without concrete advice on the approaches to conflict resolution most managers and professionals will be at a loss. A related but no less important point is that this research has been carried out based exclusively on private-sector data. This last point is addressed at length next.

Do Public-Sector Differences Affect Implementation?

There is a very legitimate and important issue for any manager or professional faced with reviewing and making use of the OPB research evidence: Are the frameworks and evidence relevant and applicable to particular circumstances they are faced with? This is particularly important for the GIS community for the following reason: the majority of the community is in the public sector, while most of the research (but not all) on OPB is based on observations in the private sector. So there are two crucial questions: First, to what extent are the results of general organizational behavior and theory (OB/OT) transferable to the public sector? Second, in particular, what can be usefully applied during the implementation of information and GIS technology in the public sector? We address these two questions to establish a means of judging the relevance of OPB as a framework grounded in largely private-sector research and its limitations.

Public Organization Theory and Behavior

The issue of public–private distinction is quite hot in some circles and has only begun to receive attention from organization theory and behavior experts as well as those in the field of public administration. At one end of the spectrum, there are those managers and professionals in the public sector that dismiss any relevance of the OB/OT because it is mainly developed in business schools to fit business managers' environment (Stevens, Wartick, & Bagby, 1988; Weiss, 1983). Golembiewski and Wildavsky (1984) characterize this as the "Dr. No" syndrome of the public agencies.

At the other end of the spectrum there are those who say "an organization is an organization"—private or public. Therefore, they see no reason for changes to the theory or prescriptions. Much of the early theory and research was based on this view (Blau, 1970; Perrow, 1967). The privatization debate is a good example of how simplistic this view has become despite the intentions of its protagonists. It is important to acknowledge that clear demarcations between the public and the private sectors are impossible. Moreover, oversimplified distinctions between public and private organizations are misleading.

This is all well and good, but we still face a paradox because researchers, managers, and professionals continue to use the public–private distinction repeatedly in relation to important issues (e.g., IT and GIS implementation) and public and private organizations differ in some obvious ways. The work of Bozeman (1987), Bozeman and Loveless (1987), and Perry and Rainey (1988) were among the first to point out the bases for the public–private distinctions and their implications.

According to their view, all organizations are public to some degree because all have some political influence and are subject to external government control, which varies based on the continuum of political and economic authority. Economic authority increases as owners and managers have more authority over the use of income and financial assets of the organization. However, it decreases as external government authorities exercise more control over these matters.

Sectoral Differences and IT Implementation Process

The majority of research on IT (in general) in the last three decades has been based on the experience of the private sector; the work of Kraemer and colleagues is the notable exception. However, as suggested earlier, the latter group has mostly concerned itself with "political impact" issues. Furthermore, there has been very little in that research that considers the validity of prescriptions based on private-sector IT research since it is largely based on public-sector evidence. More recently, this trend is starting to reverse itself (Bozeman & Bretschneider, 1986).

Using the dominance of economic/political authority framework as proposed by Bozeman, a helpful organizing scheme has been proposed by Bretschneider (1990). This scheme consists of some propositions and allows us to think through the public–private differences as far as the general IT activities (and by extension, from our standpoint GIS) are concerned during implementation. These relate to two major areas: (1) *orga-*

nizational environment (consisting of interorganizational interdependence and procedural delay [red tape]; and (2) *managerial actions* (consisting of criteria for evaluating hardware and software, and planning process). The four propositions below describe the essence of the model advanced by Bretschneider (a fifth is significant but not to our concern: the level of IT manager in the organization).

1. *Public IT managers must contend with greater levels of interdependence across organizational boundaries than do private IT managers.* The legal and constitutional arrangements in large part determine the authority of public organizations. This very political authority has a level of embeddedness in its concerns for checks and balances, which are manifested as oversight groups or external organizational control of personnel activity and financial resources as well as expectations of cooperation (noncompetition) to reduce waste. Therefore, public organizations tend to exhibit higher levels of interdependence across organizational boundaries than do private organizations.

2. *Public IT managers must contend with higher levels of red tape than private IT managers.* It is expected that greater interdependency, largely due to checks and balances (or accountability), will lead to more procedural steps for a specific management action (red tape).

3. *Criteria for the evaluation of hardware and software, which ultimately lead to purchasing decisions, are different for public IT and private IT.* There are numerous textbook and practical approaches for purchasing decisions of either hardware or software. Some reflect economic criteria such as cost–benefit analysis, while others reflect feasibility issues such as compatibility, connectivity, and the like. It is expected that differences here will reflect general differences in organizational environment and be manifested as different weights for a more or less fixed set of criteria.

4. *Public IT planning is more concerned with interorganizational linkages, while private IT is more concerned with internal coordination.* The organization behavior and theory tell us that planning is a major component of management. However, public IT management faces planning issues in a different manner. High levels of interdependency among public agencies (e.g., city departments) lead to higher levels of uncertainty and less control over the environment by any individual group. This condition leads to planning activities serving more as a vehicle for managing interorganizational linkages than coordination of effort within the organization.

Bretschneider (1990) tested these propositions on more than 1,300 public and private managers and found significant differences for all four between the public and private managers. At the more detailed level for each item the significance varied, but at the aggregate the propositions held. These results are significant for our purposes. If we can establish plausible hypotheses about the interaction of OPB and these dimensions, then our OPB framework, propositions, and prescriptions can be said to be on a far more solid basis than if they were supported solely on the private-sector data.

Bases for Organizational Politics: Six Propositions

At the outset it is important not to overstate the case for OPB, but to accurately characterize it. It is fair to assume that most organizational behavior is governed by a model that falls somewhere between the two poles of procedural rationality[1] and political behavior (Zey, 1991). In other words, both models can to varying degrees explain a particular behavior. In fact, Hardy (1987, citing Bums, 1961) has pointed out that individuals in organizations are both "rivals and cooperators," and that in a large number of cases (*but not all*) individual success is bound up with organizational success. Therefore, a realistic model of organizational behavior must account for both situations. The question then becomes as far as we are concerned: When are political factors likely to be the most important? We answer this question with six propositions.

These underscore the "logical" or "natural" view of OPB in situational contexts, where it is most likely to occur. These propositions follow a logical sequencing as they develop the argument for understanding the "true" nature of organizational politics. The objective is to show that OPB in these situations is not only not irrational or illogical, but in fact the reverse. That is, OPB grows out of certain rational and/or logical considerations of individuals and workgroups within organizations.

- *Proposition 1a: Large-scale innovations involve changes in resource allocation patterns, or reallocation of scarce resources.* According to Hardy (1987),

---

[1]The use of the term "procedural rationality" is deliberate in reference to the modified rational choice models and in recognizing that to be politically self-interested is to be rational.

political influences are likely to be particularly intense when the existing pattern of resource allocation is changing (it is important to note that this is not always the case), for such change presents opportunities to enhance power positions. Large-scale innovation decisions (like the decision to adopt GIS) typify this pattern. Furthermore, in these situations, there is a significant amount of complexity, unpredictability, and uncertainty that renders formal economic criteria for project evaluation less feasible (Wilensky, 1967, as cited in Hardy, 1987). Also, Hardy cites Gore and Dyson (1964) in support of her view: "The relations between participants in routine decisions are typically characterized by cooperation, while conflict of some sort is the norm in innovative decisions" (p. 103).

• *Proposition 1b: The organizational decision process, in the context of innovation, often involves bargaining, negotiation, and jockeying for position.* It is likely to come as no surprise to the majority of readers, particularly those who are currently employed in organizations, that the manner by which many decisions are made is often based less on purely logical decision processes than on a variety of intervening criteria. Certainly, as James March and Herbert Simon noted nearly 50 years ago, individuals strive for logic in their decision processes. However, for a variety of reasons, we are often more likely to be influenced by and to make use of a variety of extra or additional criteria in arriving at decision choices (March & Simon, 1958). One process that is common within organizations where scarce resources are the rule is to make use of bargaining or negotiation behavior. Bargaining follows one of the most common approaches to dealing with conditions of scarce resources and especially so in the context of innovations (Wilson, 1982): individuals and department heads make "deals," or compromises between the variety of competing desires and organizational reality.

• *Proposition 2a: Groups differ in terms of interest, values, attitudes, time frames, and the like, thereby making intergroup disagreements or cleavages a permanent feature of organizational life.* In 1967, a landmark study was conducted by Paul Lawrence and Jay Lorsch which sought to investigate the manner in which roles and attitudes differ among various subgroups in organizations (Lawrence & Lorsch, 1967, 1969). Through their research, Lawrence and Lorsch uncovered and introduced a phenomenon that they referred to as "organizational differentiation." The concept of differentiation was later used by Astely, Axelsson, Butler, Hickson, and Wilson (1982) to describe the fact that in certain decision situations, especially those involving strategic, structural, and technological changes, this same differentiation translates into an intensification of differences among workgroups. This contextual phenomenon is such that the workgroup behavior around these special decision topics become cleavages—that is, "semipermanent" nonconvergence of interests. However, this nonconvergence of interests, according to Bacharach and Lawler (1980), does not equate with *contention of objectives.*

   • *Proposition 2b: In the presence of high uncertainty and complexity associated with technology implementation projects, cleavages in organizations leads to workgroups exhibiting interest group or coalitional behavior.* According to Walsh, Hinings, Greenwood, and Stewart (1981: 131):

> The starting point of the political model is the existence of differentiation. An organization is conceived as being made up of separately identifiable groups differentiated both horizontally and vertically according to division of labor and authority. . . . These differentiated groups may just as easily have conflicting as coinciding interests and values.

This points out an important issue to understand: when we refer to an "organization," it may be a convenient shorthand to use the term in a monolithic sense—that is, that an organization can and will act as a single, purposeful entity. In reality, the term "organization" gives meaning to the reality behind this misperception. In both the public and the private arenas, organizations are composed of a variety of groups: labor versus management, finance versus marketing, and so forth. These groups, which must be viewed as essentially self-interested, are the sum total of what comprises an organization (March, 1962; Mintzberg, 1985; Pfeffer & Salancik, 1977; Tushman, 1977). Under certain conditions there will be pressure to act in a purely self-interested manner. In other cases, there will be impetus to act in a coalitional manner (Yates, 1985). Note also that, in effect, these first four propositions have a lot in common with the conclusions by Kraemer and his associates cited earlier. However, they are not corollaries, and in fact the scope of these propositions is thought to be far greater and more general.

   • *Proposition 3a: In the presence of complexity/uncertainty, the resource reallocation decision process of bargaining/negotiation and coalition dynamics gives rise to more intense conflict than otherwise will be the case.* This proposition forms one of the fundamental aspects of political behavior; it is the underlying rationale behind the political model of organizational life. Because of the essential cleavages (Astely et al., 1982) in certain decision

situations—innovations—as well as a higher level of engagement in negotiation and bargaining (Wilson, 1982), conflict among workgroups becomes more pronounced in organizational decision making (Bacharach &
Lawler, 1980). Therefore, the essence of the above proposition is not that
GIS implementation produces intergroup organizational conflict but
that it intensifies the process. This intensification has implications for
the decision-making processes of the organizations, which when more
conflict-riden than usual will resort to nonstandard (mostly nonco-
operative/political) modes of resolution. This is the topic of the next
proposition.

• *Proposition 3b: During innovation processes due to bargaining/negotia-
tion, coalition behavior, and the presence of conflict there is a tendency for
more intense political behavior than otherwise will the case.* On the one
hand, organizational conflict—the exertion of influence, through informal means in an intergroup organizational decision-making process—has
its origins in cleavage (Astley et al., 1982). On the other hand, organizational conflict has its roots in the resource reallocation process engen-
dered by the innovation decisions that generate disagreements (Wilson,
1982). Given the increase in the conflict level that is the result of these
two trends, the predisposition for organizational members is to resort
to political behavior in resolving these conflicts. These are the basic
propositions that constitute the framework of our view of organizational political behavior, portraying them as "rational" responses to a
situation of increased conflict. Figure 7.3 illustrates these propositions
and the underlying constructs.

Organizational Political Behavior: A Framework

Having provided the propositions that constitute the logical bases for
OPB, we now wish to establish the analytical contents of OPB. The
major goal is to distinguish between OPB and non-OPB based on the
contents of a particular behavior. After all, if every intent, action, or
outcome in an organizational situation is classified as OPB, it loses its
explanatory power. Furthermore, this is needed to set bounds on the
concept of OPB for two reasons. First, researchers must clearly delimit
the OPB construct before it can be used. Second, managers and professionals must be able to understand OPB in order to manage it and/or
effectively deal with it.

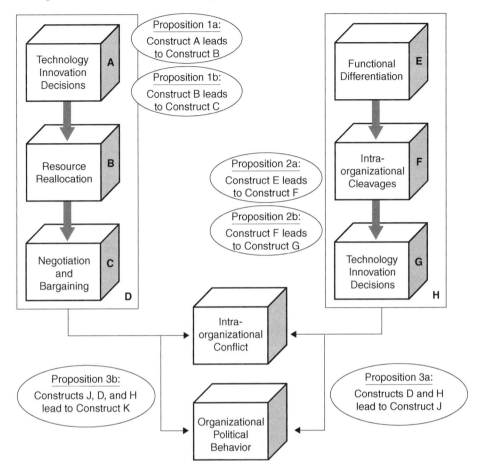

FIGURE 7.3. Constructs and propositions leading to organizational behavior. From Pinto and Azad (1994). Copyright 1994 by the Urban and Regional Information Systems Association. Reprinted by permission.

Characteristics and Components of OPB

We follow the work of Drory and Romm (1990)—in which they provide a synthesis of OPB—to present a classification of OPB contents based on certain categories. Figure 7.4 is an adaptation of their classification scheme through which we can identify OPB based on three subject categories: (1) the ends (or intent or goals or outcomes) of OPB; (2) the means employed in (or process of) OPB; and (3) the context (or situational characteristics) of OPB.

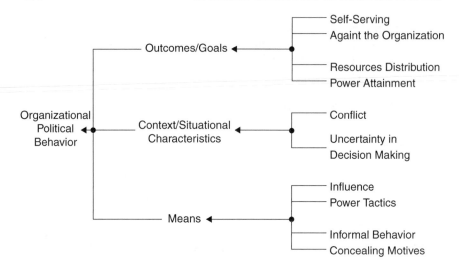

FIGURE 7.4. Contents of organizational political behavior (Drory & Romm, 1990). From Pinto and Azad (1994). Copyright 1994 by the Urban and Regional Information Systems Association. Reprinted by permission.

The typical *goals* (intent/outcomes) category of OPB according to Drory and Romm can be grouped into four subcategories:

- Self-serving
- Against the organization
- Resources distribution/redistribution
- Power attainment

The overall orientation of the outcomes/goals characterized by these various categories clearly indicates that they all deviate from formal organizational goals or even contradict them to varying degrees. The first category of goals suggests that individuals engaging in OPB are generally intent on only self-serving actions rather then organization-serving actions. The next outcome/goal category in Figure 7.4 encompasses those actions that are against the organization, implying a direct opposition of OPB to organizational goals. The remaining two self-serving OPB outcomes/goals, namely, protecting one's share in formal organizational resources and the attainment of power, are generally placed outside the formal organizational goals as well.

Drory and Romm put forward in the *means* category of OPB certain actions that are in almost all cases not endorsed by the formal organization: (1) influence, (2) power tactics, (3) informal behavior, and (4) concealing motives. For example, influence and informal behavior are often (if not always) applied in the absence of formal authority. In the case of power tactics, they are usually employed when formal rules are not sufficient. The last subcategory, concealing motives, clearly does not coincide with the formal organizational model.

It is interesting to note that the various means and outcomes subcategories seem to have yet another common characteristic. They all imply or assume to some degree the notion of potential conflict. In the majority of cases, informal influence, power tactics, and concealing motives are all employed under the assumption that the other side is not likely to cooperate under their own volition. In the case of OPB outcomes, they are also in direct or potential conflict with the formal organization and/or with other parties. Potential conflict with other organizational actors is usually a direct result of the self-serving nature of OPB intent/goals. It is axiomatic that desired outcomes that are contrary to the organization mission are by definition *in* conflict with the formal organizational goals. In addition, the attainment of outcomes relating to resource allocation and power in most cases (if not all) contradicts the interest of others. If one engages in increasing one's share in the resource allocation process or the goal of power attainment, these acts very likely will lead to conflict with other organizational members. Therefore, the element of conflict is not just a contextual (situational) characteristic that may or may not be associated with OPB; *it is at the core of any OPB situation.*

Uncertainty is another contextual (situational) characteristic that is logically derived from and structurally associated with the overall construct of OPB. That is, informal influence tactics, we suggested above, may be used more effectively where there is a lack of objective information to guide the decision-making process. Given the typical uncertainty associated with the outcomes of IT projects and GIS, one would expect more intense OPB in these situations (Mumford & Pettigrew, 1975; Pettigrew, 1973, 1975). The expected observation, therefore, is that the political actor will prefer to exploit situations of uncertainty to try and further his or her goals by using political behavior than others. However, we should qualify that observation—this does not necessarily exist in every OPB situation and as such is only optional.

In summary; when all the elements of ends, means, and context of OPB are considered together, two major common behavioral conditions (characteristics) emerge: (1) a divergence from the formal organization;

and (2) the underlying assumption of potential conflict. Thus, to formally state what is usually implicit in the discussions of the topic: *OPB is associated with organizational behavior, which deviates from the formal, techno-economic goal-oriented approach assumed by the rational models of organizations* (Drory & Romm, 1990: 1146).

As we discussed in the early part of the chapter, the term bears very negative connotations for most managers because of the core element of conflict. Stated more directly: There is no reason to believe that OPB keeps everybody happy. Although most academic definitions of the term adopt a neutral stance toward the morality or ethics of OPB, given our interest in successful GIS implementation through sensitivity to and management of OPB, we put forward a normative approach of "positive OPB." (That is not to say that the consequences of OPB may be undesirable to some individuals, groups, or to the organization at large.)

Classification of OPB

It may be noted that not all OPB situations are characterized by all the subcategories of each ends, means, and context elements. These represent an organizing framework for research and practice around the analytical contents of OPB. In fact, many situations may consist of only one or two of these categories/subcategories. The meaning of OPB in a given situation depends to a large degree on which categories/subcategories are included in the schema for analysis. A distinction can be made between three types of definitions in this respect.

Identifying OPB by Ends/Outcome

In such cases, the political behavior is defined by its goals regardless of the means employed to attain them. Defining OPB as self-serving, or as the struggle to attain power, falls within this category. This approach allows for relative flexibility in exploring behavioral tactics that might serve the purpose of attaining informal goals. It is relatively narrow in its scope, however, as it excludes all cases of manipulation and informal influence geared toward the attainment of formal outcomes. Typical collective-bargaining processes are examples of using informal means for rational and formal outcomes. Such behavior would not be considered political according to the above definitions.

Identifying OPB by Means Used

This grouping is based exclusively on the type of the means used to engage in and achieve OPB goals. The prime example of this is the exercise of power tactics. Usually, the definitions in this category refer to informal means of influence and consider the use of such means as political no matter what the nature of the anticipated outcome. Consequently, it should be clear that OPB may be applied toward the attainment of both informal and formal outcomes. Therefore, this grouping presents a relatively comprehensive view of OPB although the range of means employed may be quite specific.

Identifying OPB by Combination of Ends and Means

Work in this category is usually characterized by both means and ends, and sometimes by context, variables. However, such cases are normally limited as they tend to restrict the meaning of OPB to a particular combination of means, outcomes and conditions.

Essential Contents of OPB

Now we can draw together the above categories into a more comprehensive conceptual framework. We do not want to suggest that a single working definition of OPB is possible or workable. In fact, we agree with Drory and Romm (1990) that a multitude of working perspectives as an organizing framework may be more desirable. However, the essential analytical contents of OPB, beyond the specific definitions, can be captured through the minimal combination of the following three categories.

Influence

The presence of an element of influence in OPB is almost axiomatic. There is wide consensus in the OPB literature that political behavior is essentially influencing behavior in the sense of trying to change or affect someone's behavior or attitude. This is equivalent to Propositions 1a and 1b presented earlier. That is, under certain conditions—innovations—the use of influence (e.g., negotiation and bargaining) becomes the dominant form of decision making.

Informal Means

The use of informal means is another element implied by most definitions. It represents a divergence from the formal organizational model. Under the OPB concept, informal means may be employed in the pursuit of either informal or formal outcomes. Formal means, however, are by definition only employed for the pursuit of formal legitimate organizational outcomes. This characteristic is equivalent to Propositions 2a and 2b. In other words, under conditions of uncertainty/complexity, the use of informal means by functionally differentiated groups—for example, coalitional behavior outside the formal organizational avenues of action—becomes the primary apparatus for getting things done.

Conflict

The third essential element to the construct of OPB is conflict. The way in which conflict is derived from the very nature of the OPB means and outcomes was discussed earlier. It is therefore suggested that the presence of direct or implied conflict is immediately derived from the nature of both the means and the outcomes associated with OPB definitions. The notion of conflict was explicitly incorporated in Proposition 3a, and its presence was a key contributing factor to the OPB.

To summarize, the basic OPB situation occurs when goal attainment is sought by *informal*, rather than formal, means of *influence* in the face of potential *conflict*.

Positive Political Behavior for Successful GIS

We started by discussing how prevalent the "negative" view of OPB is. By now, we hope to have articulated a more realistic view of OPB from analytical and research angles. We want to complement these by more concrete and practical guidelines on engagement in OPB to enhance the probability of successful GIS implementation. Readers need to come to their own conclusions about their own individual views and roles in OPB. We would suggest that there are usually three distinct individual views and roles regarding OPB—two of these roles are equally inappropriate, but for entirely different reasons, and probably the major cause of the "negative" view. The first approach can be best termed the "naive" attitude regarding organizational politics. The naive view is characterized by a willingness to ignore organizational politics or simply view them as "dirty tricks" in which one resolves never to engage. Benveniste (1989), an advocate of the pragmatic approach to urban planning, has termed this view "apolitical politics" and dysfunctional (p. 72). His criticisms of the naive view of (technical) planning—and by extension GIS implementation as one such activity—is that it has the following consequences:

1. Widespread perception of mystification
2. Lack of preparation and resources to play political roles
3. Distrust by high-level executives and politicians
4. Disregard for implementation details
5. Blame of failures on politics and management
6. Distortion of (CIS) professional's role (tendency to elevate technical elegance rather than effective implementation).

The second, and opposite, approach is undertaken by individuals who enter organizations with the express purpose of using politics and aggressive manipulation to reach the top. Christie and Geis (1970), in *Studies in Machiavellianism*, put forward four characteristics of such a person:

- A relative lack of affect in interpersonal relationships
- A lack of concern with conventional morality
- A lack of gross psychopathology
- Low ideological commitment.

We refer to such people as "sharks." While actually few in number, this type readily embraces political behavior in its most virulent form. Their loyalty is entirely to themselves and their own objectives. Work with them and one is likely to be used and manipulated; get between them and their goal and their behavior becomes utterly amoral. The only cause these individuals espouse is their own. As we stated initially, we regard both the "naive" and "shark" as wrong-minded about politics, but for completely different reasons. Their attitudes underscore the awareness of the third type of organizational actor: the "politically sensible." Table 7.1 provides distinctions among characteristics of these views and roles.

Politically sensible individuals view OPB "positively" with few illusions about how major resource allocation decisions are made. Their position is characterized by factors opposite to those of the "naive" approach

TABLE 7.1. Characteristics of Political Behaviors

Characteristics	Naive	Sensible	Sharks
Underlying attitude: "Politics is . . . "	Unpleasant	Necessary	An opportunity
Intent	Avoid at all costs	Used to further department's goals	Self-serving and predatory
Techniques	Tell it like it is	Network, expand connections, use system to give and receive favors	Manipulation, use of fraud and deceit when necessary
Favorite tactics	None, the truth will win out	Negotiation, bargaining	Bully, misuse of information, cultivate and use "friends" and other contacts.

Note. From Pinto and Azad (1994). Copyright 1994 by the Urban and Regional Information Systems Association. Reprinted by permission.

but at the same time they are the antithesis of the "shark" approach. They understand, either intuitively or through their own past experiences and mistakes, that politics is simply another side, albeit an unattractive one, of the behavior in which one must engage in order to succeed in modern organizations (Pinto, 2000; Sense, 2003). The politically sensible person is apt to state that this behavior is at times necessary because "that is the way the game is played." It is also important to point out that politically sensible individuals generally do not play predatory politics, as in the case of the sharks who are seeking to advance their own careers in any manner that is expedient. Politically sensible individuals use politics as a way of making contacts, cutting deals, and gaining power and resources for their departments in order to further cooperate, rather than for personal gains.[2]

Figure 7.5 illustrates conceptually the fact that the sensible political approach is the most beneficial from the whole organization's standpoint.

---

[2]We are tempted to use the example of DigiCom, the high-tech firm in the midst of developing the "killer" virtual reality application, in Michael Crichton's novel, *Disclosure*. The characters of Tom Sanders (up-and-coming executive), Meredith Johnson (the scheming executive), and Stephanie Kaplan (shrewd but careful executive) fit surprising well our categories of naive, shark, and sensible, respectively. For those of you interested in the real-life drama of organizational political behavior and in high-technology settings, it is a fascinating book.

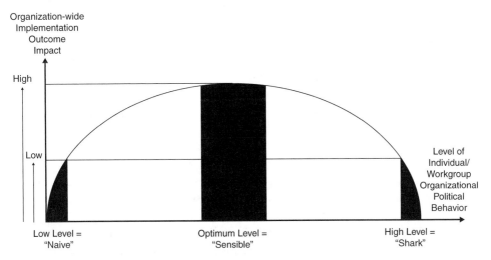

FIGURE 7.5. Conceptual representation of the relationship between level of political behavior and organizational performance. From Pinto and Azad (1994). Copyright 1994 by the Urban and Regional Information Systems Association. Reprinted by permission.

That is, assuming that in the case of some decision outcomes (say, GIS implementation), certain outcomes are to the benefit of the organization as a whole while others are either detrimental to it or less than optimal. We may represent this case by an inverted U curve which exhibits lower organizational performance at either low ("naive") or very high ("shark") levels of political behavior. However, the moderate (sensible) level of political behavior is associated with the highest level of organizational performance.

Acknowledgment of OPB

What can we summarize here? In dealing with individuals suffering from a variety of dysfunctional illnesses, therapists and counselors of all types have long taken as their starting point the importance of a patient's acknowledgment of his or her problem. Positive results cannot be achieved in a state of continued denial. While it is not our purpose to suggest that this analogy holds completely true with organizational politics, the underlying point is still important: denial of the political nature of organizations does not make that phenomenon any less potent. Organizations in both the public and the private sectors are inherently politicized for the reasons that have been previously discussed. We realize that, in offering this view, we run the risk of offending some readers who are uncomfortable with the idea of politics and believe that, somehow, through the combined efforts of all organizational actors, it is possible to eradicate the political nature of companies or governmental agencies. We must disagree, as will, we believe, the majority of managers in organizations today. Politics are too deeply rooted within organizational operations to be treated as some aberrant form of bacteria or diseased tissue that can be excised from the organization's body.

The first implication argues that before managers are able to learn to utilize politics in a manner that is supportive of GIS implementation, they must first acknowledge (1) their existence, and (2) their impact on adoption success. Once we have created a collective basis of understanding regarding the political nature of organizations, it is possible to begin to develop some action steps that will aid in GIS implementation.

Doing Your OPB Homework: Stakeholder Analysis

Another way to illustrate the essential conflict in GIS implementation is through the use of stakeholder analysis. Stakeholder analysis is a useful tool for demonstrating some of the seemingly unresolvable conflicts that occur through the planned introduction of any new information system such as GIS. The concept of an organizational "stakeholder" refers to any individual or group that has an active stake in the success or failure of the planned GIS. For example, top management, as a group, once committed to acquiring a GIS, has an active stake in that GIS being accepted and used by client departments. For the purposes of simplicity, we are categorizing together all members of upper management as one stakeholder group. A valid argument could be made that there are obviously differing degrees of enthusiasm for and commitment to the adoption of GIS technology (Azad, 1997). In other words, a good deal of conflict and differences of opinion will be discovered within any generalized group. Nevertheless, this approach is useful because it demonstrates the inherent nature of conflict arising from GIS adoption as it exists *between* stakeholder groups, rather than *within* them.

For the sake of our discussion, let us assume that in the case of an effort to implement a new GIS, there are four identifiable stakeholder groups: top management, the accountant, the clients, and the manager's own implementation team. As we suggested previously, "top" management has given the initial go-ahead to acquire and install the GIS. Likewise, the accountant provides the control and support for the implementation effort, ensuring that budgets are maintained and the project is coming in near projected levels. The clients are the most obvious stakeholder as they are the intended recipients of the new system. Finally, assuming an implementation team is working together to implement the GIS, this team itself has a stakeholder interest in the implementation, particularly if they are receiving some type of evaluation for their efforts.

To demonstrate the nature of conflict among stakeholder groups, we have developed three criteria under which the implementation will be evaluated: schedule, budget, and performance specifications. "Schedule" refers to the projected time frame to complete the installation and get the system online. "Budget" refers to the implementation team's adherence to initial budget figures for the GIS adoption. Finally, "performance specifications" involve the assessment that the GIS is up and running, while performing the range of tasks for which it was acquired. Certainly, additional evaluative criteria can and should be employed; however, for simplicity's sake, these three success measures serve to illustrate the nature of the underlying conflict in any GIS implementation.

Figure 7.6 shows the four identified stakeholders and the three success criteria that have been selected.[3] The arrows are used to illustrate the emphasis placed on each of these criteria by the stakeholders. For example, consider the case of stakeholder preferences in terms of the differences between clients and the implementation team. It is obvious that in

	Cost	Schedule	Performance Specs.
Top Management	↓	↓	—
Accountant	↓	—	—
Client	↓	↓	↑
Implementation Team	↑	↑	↓

FIGURE 7.6. Stakeholder analysis. From Pinto and Azad (1994). Copyright 1994 by the Urban and Regional Information Systems Association. Reprinted by permission.

---

[3]We are grateful to our colleague Bob Graham for this insight.

terms of evaluation criteria such as schedule and budget, there are significant differences in attitude: the clients want the system delivered as soon as possible for as cheap a final price as possible. On the other hand, the implementation team would like large budgets and longer installation schedules because that takes the pressure off the team in terms of bringing the system online. Furthermore, the criterion of performance specifications will vary by stakeholder group. Clients want the opportunity to alter the system, customize it, or add as many technical capabilities as possible. The implementation team is much more comfortable with a simple system that has few technical surprises (and therefore less likelihood of long debugging procedures) and is not changed or modified once it has been acquired.

Figure 7.6 presents a compelling case for the underlying conflict of most GIS implementation efforts. It also serves to illustrate one inescapable conclusion: in order to rationalize and resolve the varied goals and priorities of the various stakeholders, a considerable amount of bargaining and negotiation is called for. As the reader will recall, bargaining and negotiation are two of the primary defining elements of organizational politics. Clearly, political behavior is required in successful implementation efforts. If we take as our starting point the conclusion that a successful GIS manager is not one who will satisfy all stakeholder parties, it becomes clear that implementation success is instead predicated on the GIS manager's ability to successfully bargain and negotiate with the various stakeholders in order to maintain a balance between their needs and the realities of the GIS implementation process. Implementation becomes a process that depends on the GIS manager's clever and effective exercise of political skills.

The one important implication of this discussion of project managers tasked with implementing a GIS is the necessity of cultivating the ability to use organizational politics effectively. By "effectively," we do not mean to imply that politics should be practiced in predatory ways. In fact, that approach is likely to seriously backfire on the viability of GIS managers who need to develop trust and goodwill to implement their systems. Rather, they must learn to appreciate and use politics as a negotiating and bargaining tool in pursuit of their ultimate goal of installing the GIS. This is another example of the reason we had earlier suggested that both the politically "naive" and the "shark" are equally inappropriate personae for managers to adopt. Successful implementation will not occur without the use of political behavior. Conversely, however, the degree of rancor and bitterness that is usually a by-product of predatory political behavior is one of the surest ways to torpedo the introduction of a new system.

Active Engagement in OPB

An understanding of the political side of organizations and the often intensely political nature of system implementation gives rise to the concomitant need to develop appropriate attitudes and strategies that help GIS managers operate effectively within the system. What are some steps that GIS managers can take to become politically astute, if this approach is so necessary to effective GIS implementation?

Learn and Cultivate "Positive" OPB

This principle reinforces the earlier argument that, although politics exists, the manner in which organizational actors use politics determines whether or not the political arena is a healthy or unhealthy one. We have tried to assert (see Table 7.1) that there are appropriate and inappropriate methods for using politics. Since the purpose of all political behavior is to develop and keep power, we believe that both the politically naive and the shark personalities are equally misguided and, perhaps surprisingly, *equally* damaging to the likelihood of GIS implementation success. A GIS manager who, either through naiveté or stubbornness, refuses to exploit the political arena is destined to not be nearly as effective in introducing the GIS as is an implementation manager who knows how to use politics effectively—that is, to promote the organization's overall goals which include the development and use of geographic information technologies. On the other hand, GIS managers who are so politicized as to appear predatory and aggressive to their colleagues are doomed to create an atmosphere of such distrust and personal animosity that there is also little chance for successful GIS adoption.

Pursuing the middle ground of political sensibility is the key to new system implementation success. The process of developing and applying appropriate political tactics means using politics as it can most effectively be used: as a basis for negotiation and bargaining. As Table 7.1 pointed out, politically sensible managers understand that initiating any sort of organizational change, such as installing a GIS, is bound to reshuffle the distribution of power within the organization. That is likely to make many departments and managers very nervous as they begin to wonder how the future power relationships will be rearranged. "Politically sensible" implies being politically sensitive to the concerns of powerful stakeholder groups. Legitimate or not, their concerns about the new GIS are real and must be addressed. Appropriate political tactics and behavior include making alliances with powerful members of other stakeholder departments, negotiating mutually acceptable solutions to seemingly unsolvable problems, and recognizing that most organizational activities are predicated on the give-and-take of negotiation and compromise. It is through these uses of political behavior that managers of GIS implementation efforts put themselves in the position to most effectively influence the successful introduction of their systems.

Understand, Accept, and Practice "WIIFM"

One of the hardest lessons for newcomers to organizations to internalize is the consistently expressed and displayed primacy of departmental loyalties and self-interest over organization-wide concerns. There are many times when novice managers will feel frustrated at the "foot dragging" of other departments and individuals to accept new ideas or systems that are "good for them." It is vital that these managers understand that the beauty of a new GIS is truly in the eyes of the beholder. One may be absolutely convinced that GIS technology will be beneficial to the organization. However, convincing members of other departments of this truth is a different matter altogether.

We must understand that other departments, including system stakeholders, are not likely to offer their help and support of the GIS unless they perceive that it is in their interests to do so. Simply assuming that these departments understand the value of a GIS is simplistic and usually wrong. Our colleague Bob Graham likes to refer to the principle of "WIIFM" when describing the reactions of stakeholder groups to new innovations. "WIIFM" is an acronym that means "What's In It For Me?"[4] This is the question most often asked by individuals and departments when presented with requests for their aid. They are asking why they should help ease the transition period to introducing and using a new system. The worst response GIS managers can make is to assume that the

---

[4] In fact, the title of a book by Robin Woods on the concept of partnering with internal and external customers is *What Is In It for Me?* (New York: American Management Association, 1993).

stakeholders will automatically appreciate and value the GIS as much as they themselves do. Graham's point is that time and care must be taken to use politics effectively, to cultivate a relationship with power holders, and to make the deals that need to be made to bring the system online. This is the essence of political sensibility: being level-headed enough to have few illusions about the difficulties one is likely to encounter in attempting to install a new system such as GIS.

GIS Implementation and OPB: Two Illustrative Cases

The use of GIS in the transportation business, and particularly for state departments of transportation, is gaining momentum. In part, this growth is fueled by the Intermodal Surface Transportation Efficiency Act (IS-TEA) of 1991 (the federal government has mandated information requirements on these agencies that would be difficult if not impossible to meet without the use of GIS technology). Most of these agencies have embarked on some form of stand-alone (bottom-up) GIS implementation. However, subsequent technological difficulties and especially organizational problems have generated enough dissatisfaction that most, if not all, of these agencies have engaged in organization-wide GIS strategic plan development and implementation. Using two such cases—pseudo-named XDOT and YDOT because of nondisclosure agreements—we will highlight some of the major GIS implementation steps and their possible political nature—especially, how the political managerial actions are applicable (or not). We will advance our interpretation of how the agencies have dealt with (or are dealing with) these situations in politically productive or nonproductive manner.

General Implementation Steps

Implementation of technologies in organizations has been the subject of investigation by researchers and practitioners for more than a quarter of a century. There are many ways of characterizing the process: by phases, by activities, by episodes, by actions, by steps, and so forth. Each author seems to have a favorite. Obviously, the choice of a particular scheme has to do with what one is trying to use the elements of the implementation process for. For example, Antenucci, Brown, Croswell, and Kevany (1991) adopt the GIS steps all in sequence. Huxhold (1991) also proposes a similar GIS-stages scheme, while Montgomery and Schuck (1993), focusing on spatial data conversion, suggest a phase approach.

For studying OPB in the implementation process, we propose a variant of these approaches that we call "tracks of the implementation process." The object is to give some (artificial) structure to the implementation process for identifying events of interest that involve major *decision points* in the life of a GIS project. The advantage of tracks is in letting us classify implementation events in sequence and parallel for the purposes of the study without imposing any rigid framework on the actual events taking place (i.e., it is purely a cognitive device). A typical GIS implementation process can have six tracks (or more depending on one's scheme): organization, application, database, systems, training/education, and funding. The typical contents of each track will naturally vary. For example, some of the key steps in the organization track would include obtaining a mandate, establishing an executive committee, appointing a statewide coordinator, and so forth. Likewise, the education track may include such activities as producing an executive summary, developing a newsletter, and conducting annual workshops. Potentially, all tracks with decision points can be contentious, although in reality that is not the case for all situations.

Case 1: XDOT

XDOT is a large state department of transportation (SDOT). It has over 13,000 employees, and is responsible for the upkeep of 44,000 miles of roads as well as a variety of other transportation mode facilities. However, the majority of its work is related to roads: 11,000 of the 13,000 employees work in road-related activities. The general organizational climate or culture can be characterized as very formalized (Burns & Stalker, 1961).

XDOT has been using CAD technology since the early 1980s and is considered one of the leading SDOTs in this regard, with over 300 CAD workstations. Beginning in the late 1980s, several divisions had shown interest in GIS, the main one being the paper-map production division. It was natural that staff saw GIS as an extension of what they were already doing. (This division is housed in the planning directorate of the XDOT, which also houses all information reporting for transportation system performance mandated by the federal government.)

In 1988, experimentation with GIS began through several pilot projects. Ultimately, this work led to the hiring of a consultant group to develop a GIS strategic plan. This plan called for a five-phase implementation strategy (and provided projected costs for each phase): strategic plan development, detailed implementation plan, core spatial database construction/application development, database construction/application development expansion, and agencywide distribution. XDOT proceeded with the plan and hired another consultant for the development of the detailed implementation plan (hereafter just the Plan—one of the authors was a senior member of the consultant team directly taking part in the Plan development process). The activities of the Plan development contained the major elements of the tracks presented earlier in approximately the same sequence.

The project was overseen out of the planning directorate of the agency which was associated with GIS activities since their inception. However, as one of the recommendations of the strategic plan, the GIS steering committee was the final (formal) arbiter of Plan contents during major disagreements. The major track events of interest, due to expected conflict in decision making, were the following:

- Organizational location of GIS unit and hiring new staff
- The mechanism for moving (attribute) data from and to division databases
- Standards for GIS application development
- Scale of the base map
- Geographic features priorities for databases and application development priorities
- Standard software and hardware platform selection and procurement
- Job reclassifications to GIS-based occupations.

Organizational Location of the GIS Unit

The staff of the planning directorate were quite keen on having the GIS unit located in their workgroup. This, however, was rarely mentioned explicitly, and assumed to be the consultant recommendation based on historical reasons and pilot application development. On several occasions when the point was brought up by the consultant team, it was subtly evaded and regarded as unnecessary for inclusion in the consultant report (the consultant team was aware that at other sites this is a point that can stifle GIS implementation and must be explicitly dealt with).

The staff of the planning directorate expressed worry about the topic in the following manner: if options other than those of the planning directorate are considered, then either the MIS division or the CAD division will try to "steal" the GIS unit. The rationale being that either had more experience with IT through standard MIS operations or the CAD use. The consultant team considered other location options and recommended that the GIS unit stay with the planning directorate for a period of up to 3 years, at which time a reevaluation of the location decision can be made. The planning directorate staff at first were not happy with the explicit location options assessments. However, they accepted it based on its short-term recommendations and other factors that are explained below. This appears to be a very contentious point in most GIS implementations, and it was so at XDOT. However, a mechanism for its resolution can be highlighted through the next decision point.

Hiring New GIS Staff

For historical reasons XDOT has been resistant to hiring new staff and, in fact, turnover rates for technical staff are somewhat low—under 10%. Perhaps the state being the major employer is an explanation for the low rate. In any event, as a part of the strategic plan recommendations and the Plan recommendations, three new hires were proposed. The agency had gone ahead with one new hire for GIS—a recent graduate with excellent programming and web management skills and some GIS training.

However, two other staff members were "loaned" out of the MIS division to the GIS unit. This produced some friction for three reasons: the staff members were "old timers" with little or no workstation know-how; their education dated back some years; and most important they were not under the direct control of the GIS unit (MIS still signed their time sheets). Initially, one of the senior managers in the planning directorate expressed much concern over management of "loaned" staff and its implications for the real GIS priorities. Finally, in a later meeting that same manager appeared to have accepted the situation as a fact of life and more to the point as a quid pro quo for the head of the administration directorate allowing the GIS unit to remain in the planning directorate (this is an inference made by the authors because such comments were never explicitly made, but always indirectly; however, it also serves to highlight typical problems of doing research on OPB given its informal nature in the majority of cases).

Mechanisms for Moving Attribute Data from MIS to GIS

XDOT has one of the most highly developed road information systems in the country. However, like the majority of these systems, it was developed in the mid-1980s and it is highly dependent on a legacy system that cost millions of dollars to develop. As such, it poses enormous and expensive technological integration problems for GIS, not to speak of organizational and cultural integration issues.

These technological and cultural clashes manifested themselves in the form of the MIS division being very resistant to providing any kind of "live" access to their attribute road data (a GIS life blood) for the GIS unit. They expressed their opposition in the form of concern over three issues:

- Being burdened with writing custom code for extraction of data to specific GIS requirements
- Concern over data integrity and quality issues, if live access is provided to someone "who does not understand the mainframe"
- Risk to data not just on the road information system but on other parts of the system with more sensitive information such as driver's license data.

Some of their concerns over security and integrity were very legitimate and understandable. However, their disagreement over live access was not totally warranted. After several meetings with the consultant team and explanations of the need for eventuality of live access, they evolved toward a softer position. They agreed to a three-phase strategy of one-way access (no live or "dead" updating of data or no two-way links, only read access). This, in the minds of the GIS staff, was a major disappointment given that the major touted benefit of GIS by vendors is its ability to provide attribute feature update through geographic feature manipulation.

The MIS division finally agreed to the consultant team's recommended phased solution. The solution would entail three phases. In phase one, they would be creating a shadow database that would provide access to the data but it would be updated every 2 to 3 months, and the communication would be only one-way to GIS as a user of data. In phase two, some "live" one-way access will be provided through the so-called pipeline technology, while the rest will still be duplicated in the shadow database. In the third phase, the link will be fully "live" but still with one-way access. The issue of two-way access was never resolved and was left to a later stage in the process of implementation.

Standards for Application Development

Since the beginning, the MIS division had a very intimate involvement in the GIS procurement process. Initially, the MIS had approached the GIS implementation as just another application and therefore wanted to subject it to the rules governing the MIS application development process. However, this implied either MIS or their consultants would design and develop the applications or an intense period of learning all the rules and methods by the GIS unit. Almost all data processing know-how was concentrated in the MIS division. In addition, it turned out that most other divisions that had tried to grow their own mini-MIS had either "failed" or were "presumed as failures" by the MIS division.

This created a significant bone of contention between the GIS unit/consultant team on one side and MIS on the other. After several presentations to the GIS steering committee as well as individual meetings with MIS management, the MIS appeared to accept GIS as new technology whose development process may be governed by different rules—at least at the beginning. This position, however, seemed to have been somewhat related to the MIS's insistence on no two-way links to GIS for some time into the future (again a quid pro quo).

Base-Map Scale

One of the key tasks in the Plan development was determining the most suitable base-map scale and providing recommendations for it. This process is often rife with disagreements over what each function within an agency is apt to call their appropriate GIS base-map scale. However, two issues usually are overriding: the first series of GIS applications and their scale requirements, and the costs associated with developing the base map at a particular scale. It goes without saying that as scale becomes larger and accuracy increases, costs go up geometrically. Therefore, one would expect major conflict and disagreements before a choice is made.

However, at XDOT this turned out to be something of a nonissue. Within the GIS steering committee, there was consensus that application development should proceed based on the updated version of XDOT's state centerline file at the 1:24,000 scale and at the accuracy of ±50 feet.

The authors can only conjecture, based on interviews with CAD engineers, that CAD work with its project and small-area focus requires the large scales and high levels of accuracy. And that is quite different from the GIS uses foreseen by the plan.

Geographic Features Database Design and Construction (and Application Development) Priorities

A large number of interviews were conducted to uncover the user needs for particular geographic information features as well as the related attribute information. In fact, the interview list was expanded midway in the project based on issues raised by various division heads and section chiefs in one of the GIS steering committee meetings. However, the expansion of the interview list did not appear to change the database construction list, which appeared largely governed by two external factors: a general concern in XDOT and state government about highway safety, making such application a first priority; and Intermodal Surface Transportation Efficiency Act (ISTEA)–mandated management systems requiring particular information about priorities already identified. It appears as though the interview list was expanded to input and change the priority list of database construction. However, there was no observed change in the priority list (perhaps a means of coopting opposition, etc.).

Standard Software/Hardware Platform

It is customary for agencies acquiring GIS technology to spend significant effort and money to "benchmark" vendors of hardware and software before a procurement decision is made. Of course, the legal aspects of procurement as well as government regulation regarding competitive bidding, and the like has a lot to do with this. XDOT was no different, and such work was stipulated in the detailed implementation plan. As the consultant team started working on the plan development, however, the client appeared very keen on staying with their current hardware and software vendor (for both CAD and GIS). In fact, several indirect allusions were made to the effect that, although there is a formal review on the benchmarking part, the GIS unit had a strong preference for one vendor.

In several other discussions, a consultant team suggested that if the client is not seriously interested in the benchmarking, the effort can be reallocated to another task. The client agreed. In the next GIS steering committee meeting the suggestion for dropping the bench-marking task and adding other tasks was formally presented by the consultant team. All XDOT participants agreed, except for the representative from the Governor's Office on Information Technology. The representative insisted that she had not heard any "good" reason for dropping benchmarking. It appears as though XDOT is favoring vendors, which is a practice not advisable in state government. The XDOT staff gave the go-ahead for the benchmarking and the task was conducted. The results, however, were largely superfluous because the consultant team already was aware of a large procurement package for the existing vendor's products.

There appeared to be an implicit element of appeasement in the informal exchange between the governor's office (and their concern for the appearance of impropriety) and XDOT (and their concern for expediency).

Job Reclassification Issues

Two separate but related matters of job reclassification appeared to surface during the Plan development. First, the human resource division of XDOT was resistant to and not convinced of developing GIS-related job categories. Their rationale was that CAD was introduced into the agency, and the engineers who were using CAD were still classified as engineers even though occupationally they used CAD in their work; in their view, what is the difference in GIS? The consultant team provided several examples of GIS job reclassifications from other agencies (largely non-SDOTs). The team also explained the need for a career path of GIS specialists, otherwise the agency may be faced with severe turnover problems if these specialists realize there are no career paths for them in a transportation agency. The human resource staff usually countered with the belief that the informal culture of an SDOT, such as XDOT, where engineering careers are the route to upward mobility, will not value such a career path. At the close of the Plan development, this issue was still not resolved.

Second, there was the issue of existing jobs being reclassified. In particular, an interesting incident currently under grievance arbitration at XDOT highlights this. The professional cartographers in the paper-map production division had filed a request for being reclassified as computer analysts with greater pay and better career prospects—with intent to sue if denied, since they were unionized. Their rationale was that their work has become similar to a computer analyst in content due to work with CAD and GIS equipment, and therefore they should be reclassified and upgraded in terms of occupational category. This incident happened just before the Plan was developed. It is now in arbitration. It shows a potential conflict that must be addressed in the long run if the GIS profession is to flourish. That is, successful implementation of GIS requires people with complete career paths moving in and filling the GIS positions in organizations.

Case 2: YDOT

YDOT is a small state department of transportation, with over 700 employees and close to 6,000 miles of roads to maintain. YDOT is also involved with other transportation modes, but much like other SDOTs, its major function is to construct and maintain highways. The agency has been using CAD in its design and construction work. It has also been developing major highway information systems since the mid-1980s. The major information system platform in the MIS division is mini-computer-based. The general organizational climate or culture can be characterized as "loosely coupled" (Weick, 1978).

YDOT engaged in a major information engineering effort in the late 1980s to early 1990s and developed an agency-wide transportation information plan. The plan produced more than 3,000 fields and 143 tables of information. Shortly following that project, there was a GIS implementation effort for which a consultant team was hired to give advice on the GIS effort within YDOT. However, the team was not hired by the workgroup that was building the GIS, but by the MIS division. At the outset, it should be made clear that consultant advice was not heeded by the GIS workgroup, perhaps understandably, as it was considered hostile advice.

In fact, the project was terminated a year after the consultant was hired because YDOT did not meet federal reporting deadlines. Subsequently, the consultant team was rehired and is in the process of developing an agency-wide GIS implementation plan. What follows is the case history of the implementation, including hardware architecture, base-map features and scale, application development linkages, and standard hardware/software vendor selection.

Hardware Architecture

An issue within the GIS community revolves around the appropriate hardware architecture for GIS. Specifically, what has been the migration rate from workstation to PC-based systems? The answers are usually based on preference rather than objective tests. However, the conventional wisdom is that, for large databases of 1 gigabyte or more in a networked environment, the platform of choice is the workstation, because its performance is still somewhat more robust than PC-based networks.

This issue also was a point of contention within the YDOT GIS effort. The GIS people went ahead and attempted to implement a PC-networked GIS. Others, especially those maintaining the paper-map production capabilities, were more interested in workstation technology. But funding from federal agencies allowed the workgroup to bypass the formal organizational procedures normally required for such projects. Later, as we discuss the issue of map scale, the hardware platform choice may have been flawed from that perspective, if not from others.

Base-Map Features and Scale

As mentioned in the XDOT case, the decision process associated with deciding on a GIS base-map scale is rife with conflict. The YDOT case, in fact, did exhibit this behavior (as opposed to XDOT, which did not). The GIS workgroup decided to adopt a very large scale—1:100 for a parcel-level map, and TIGER (1:100,000) for their centerline file. This appeared contradictory to a number of the YDOT players, since the prevalent base-map scale for an SDOT is the centerline file with a 1:24,000 scale. Also, maintaining parcel-level base maps at 1:100 scales requires terabytes of storage per county. One can only imagine the storage and software/hardware architecture requirements. The choice for scales appears not well thought out, again driven perhaps by the funding availability through external sources.

Application Development Linkages and Priorities

The information engineering effort had identified a pavement management system (PMS) as an application priority, while GIS technology was regarded as just another application. In addition, given the behavior of the GIS unit, the impetus was there for the PMS to be developed independent of the GIS, since the GIS workgroup appeared not to be in accord with the rest of the agency's needs and priorities. In any event, a separate pavement management system was developed by a workgroup. However, it did not use state-of-the-art GIS technology methods such as dynamic segmentation, which is a real boon to PMS development (the state of Wisconsin Department of Transportation has been the leading SDOT in demonstrating this application). Although one cannot put the blame for that on the GIS workgroup, the fact that agency-wide considerations were not given top priority meant that efforts such as PMS and GIS were "islands of information and know-how" with little integration. In addition, the GIS workgroup application priorities were not the result of any agency-wide user needs assessment, and it did not even correspond to the information engineering (IE) effort. Consequently, the rest of the agency did not take much interest in ensuring the successful completion of the project.

Standard Hardware/Software Platform Selection

The GIS workgroup did not go through an evaluation procedure, and decided to adopt a PC platform as well as a desktop GIS package. This was at odds with the purchasing decisions of the rest of the agency, especially the paper-map production unit, which was using workstations and their own software vendor. The same platform was used by the CAD people across the agency. Because the PMS's application architecture was closed and proprietary, the agency relies on a different vendor to maintain it, and conversion to a multiuser system has been even more costly and lengthy (this application does not use an off-the-shelf GIS package).

The GIS workgroup's decision appears not to be motivated by the agency-wide considerations of know-how and compatibility, but rather dictated by the workgroup's interests. Again, the outside funding of the project allowed the GIS workgroup to bypass normal channels of decision making and adherence to organization-wide goals and mission.

The GIS project did not meet the deadline for the EPA regulatory filings. This resulted in a major delay in the award of the state's federal transit grants which stipulate that EPA requirements have to be satisfied. Subsequently, the project funding was cut to zero, staff were reassigned, and in particular, the division manager was demoted to a position in another division. This failure can be attributed to a number of factors, but most important was that this effort was regarded as a "rogue elephant" in the mind of YDOT's other players. Perhaps it is a classic conflict situation—a workgroup attempting to implement new technology by "going it alone," sometimes against the organizational goals (almost the replica of what was discussed in terms of content of OPB), and then not succeeding.

It is interesting to note that, for this SDOT, the new GIS Plan development now under development has coincided with a change of administration in state government.

Reflections on OPB in GIS Planning and Implementation

In each SDOT case we have elements of behavior in the organization that can be characterized as political. However, GIS technology acquisition and implementation in each case is not just one decision. It is a process of many decisions and events. It involves a number of subcomponents, namely, "track events." Therefore, given the details of these cases and the framework presented earlier for OPB, we can speculate on the "validity" of (1) the logical propositions and constructs, (2) the analytical contents of OPB, and (3) the types of political actions taken by the actors in each case.

Constructs and Propositions

A summary of our analysis of the cases based on propositions and constructs appears in Table 7.2. Regarding Proposition 1a, namely, that innovation decisions lead to resource reallocations, we have strong support as far as the XDOT case is concerned. The organizational location decision for the GIS unit was a clear example of resource reallocation that was supported by the planning directorate. This was thought to have brought with it extra resources and prestige along the lines of Kraemer and colleagues.

In the YDOT case, there was consensus among key actors in the organization from other divisions that the GIS unit was interested in developing the GIS unit's goal primarily to "command" a first place in the organization regarding GIS expertise and respect.

Negotiation and bargaining as hallmarks of innovation-induced resource rereallocation was expressed through Proposition 1b. The "loaning " of, versus the hiring of, new staff in the XDOT case pointed to this kind of behavior. Especially, the quid pro quo between the heads of the planning directorate and the management and budget directorate regarding the "exchange" of a GIS unit location for only one new hire and the use of existing MIS staff indicates this kind of negotiation and bargaining.

In the YDOT case, such behavior was not readily observable. However, this was partially due to lack of researcher involvement at the early stages of the project (and the external funding of the project). Therefore, given that the claim on internal resources was largely untouched, it would be expected that we will not see as much bargaining and negotiation as would be the case if the project was funded by internal monies.

Proposition 2a related the functional differentiation to cleavages among organizational workgroups in the context of an innovation deci-

TABLE 7.2. Assessment of Organizational Political Behavior Constructs and Propositions based on XDOT and YDOT Cases

OPB proposition	XDOT	YDOT
1a. Innovation decisions lead to resource reallocation	(+) Organizational location decision was a clear example of this	(?) Not directly observed because of the use of outside funds
1b. Resource reallocation decisions are characterized by negotiation, bargaining, and jockeying for position	(+) Hiring new staff versus "loaning" staff indicated behind-the-scenes negotiations	(?) Not directly observed because of the use of outside funds
2a. Under uncertainty/ innovation conditions, functional differentiation leads to major intergroup cleavages	(+) MIS vs. GIS disagreement over attribute-data transfer to GIS	(+) GIS unit made decisions regarding base map scale, hardware, standard software, etc. in self-interested manner
2b. Intergroup cleavages lead to workgroup coalitional behavior	(+) MIS with the help of management/budget directorate formed a coalition to block the short-term "live" attribute-data transfer	(+) As above
3a. Bargaining/ negotiating, and coalitional behavior, leads to interpersonal and intergroup conflict	(+) New staff, choice of platform, attribute-data transfer, etc. led to GIS and other groups being at odds or in conflict	(+) Intergroup conflict was more apparent after the GIS project was terminated based on opinions of other players
3b. Bargaining/ negotiation, coalitional behavior, and conflict lead to OPB	(+) All of the above as well as job reclassification lead to relatively "positive" level of OPB	(+) GIS group's OPB was viewed by most other players as "dysfunctional" (or shark-like)

Note. (+) = behavior present; (–) = not present; (?) = inconclusive evidence. From Pinto and Azad (1994). Copyright 1994 by the Urban and Regional Information Systems Association. Reprinted by permission.

sion like GIS. In the XDOT case, the MIS division clearly engaged in self-interested behavior protecting their turf as having full control over attribute-data resources. Although the long-run ability for "two-way" update of attribute data is in the XDOT organization-wide interest in productivity, it was resisted by the MIS division. It succeeded in postponing a decision on this issue until a later date (if at all!).

In the case of YDOT, the decision on scale of base map, hardware platform, and standard software for GIS by the GIS unit were made unilaterally, without regard for other divisions' needs. The self-interested behavior of this workgroup was based on opinions expressed by other workgroup members that did not share the priorities of the GIS unit, and who were already using another hardware/software platform.

The coalitional behavior—contained in Proposition 2b—was clearly demonstrated in the XDOT case. The MIS division was able to control the outcomes of the GIS Steering Committee meetings on this topic, and managed to effectively block the "two-way" update of attribute data. They accomplished this by clearly allying themselves with the management and budget directorate, which saw itself as the "custodian" of all electronic attribute data, and did not allow encroachment, viewing it as either generating more work for them and/or loss of control of data.

In the YDOT case, the GIS unit again acted unilaterally as an interest group with little attention to other workgroup needs in almost all track events regarding hardware, software, base map scale, and so on. The opposition of the rest of the organization became more apparent after the project was terminated. This workgroup is not getting cooperation from other workgroups in designing some critical information systems for the federal reporting requirements. In addition, their GIS is looked at as incompatible with any new organization-wide GIS effort.

The existence of conflict—expressed in Proposition 3a—is almost axiomatic in all political situations in organizations. Nevertheless, it is important to explicitly reference them. For the XDOT organization, the new staff hire, GIS unit location, and attribute-data transfer all led to the GIS unit and other groups—particularly the MIS division—being in conflict. The conflicts were manifested as mild to intense disagreements. They never reached dysfunctional levels, except in the case of attribute-data transfer. Even in that case, the GIS unit, with the help of the consultants, managed to develop an interim solution that could demonstrate the advantages of two-way attribute-data transfer.

The existence of conflict in YDOT during the GIS implementation manifested itself through several decisions regarding base-map scale as well as hardware/software choice. The disagreements were particularly intense when consultants were brought in to mediate or to present contrary opinions to the GIS unit's unilateral decisions.

For each case, different levels of OPB were operating. In the case of XDOT, the OPB was based on the content factors of influence, informal means, and situations of conflict. In the YDOT case, these factors were not as readily observable during the implementation process, although they became more intense as the outside consultants (the authors) were brought in. However, the behavior in the posttermination phase of the project has clearly demonstrated classic OPB factors: informal means, influence and conflict.

OPB Content in XDOT and YDOT

In analyzing events of a case in hindsight, there is always the danger of seeing the events in the way the framework favors. However, case studies highlight the favorable and not-so-favorable aspects of this framework.

In the XDOT case, we can say with relative certainty that almost all decisions involved high levels of self-interest, resource redistribution, power attainment, and informal behavior. On the other hand, with the exception of the attribute-data transfer, working against the organization and concealing motives was rare.

Because of less author involvement with the YDOT case, the same generalization is more risky. However, it is possible to say that activities counter to organization-wide interests were more pervasive in the YDOT case in almost all decisions. See Table 7.3 for our speculations on the presence of these factors in both cases, and their intensity.

The objective of this analytical speculation is not to test our OPB contents—that would be beyond the scope of the current chapter—but to

TABLE 7.3. Assessment of Organizational Political Behavior during GIS Implementation at XDOT and YDOT

OPB category	OPB subcategory	XDOT	YDOT
Outcomes	Self-serving	+ + +	+ + +
	Against the organization	+	+ + +
	Resources (re-) distribution	+ + +	+ + +
	Power attainment	+ + +	+ + +
Means	Influence	+ + +	?
	Power tactics	+ +	?
	Informal behavior	+ + +	+ + +
	Concealing motives	+	?
Situational characteristics	Conflict	+ +	+ + +
	Uncertainty	+ +	+ + +

Note. Activity of that type was estimated to be occurring at the following levels: ? = difficult to assess; + = low; + + = moderate; + + + = high. From Pinto and Azad (1994). Copyright 1994 by the Urban and Regional Information Systems Association. Reprinted by permission.

show it is possible to readily observe OPB. The various factors in GIS implementation need to be more fully explored to assess their reliability and validity.

Assessment of OPB Based on Managerial Actions

We have demonstrated the presence of organizational political behavior during GIS implementation in XDOT and YDOT. However, one of the key motivations of this chapter was to present certain "guidelines" for GIS managers and professionals that will help them in their projects. These guidelines included doing a stakeholder analysis, but more important, to cultivate a positive OPB culture, as well as practice of "WIIFM" (or engagement through exchange). It is imperative after having presented these ideas to assess the extent these practices were observed in the XDOT and YDOT cases.

Table 7.4 is a distillation of this assessment for both cases. We have listed the major GIS implementation track events and provided our assessment of OPB culture and/or practice of "WIIFM." In our assessment, the XDOT and YDOT GIS implementation cases demonstrate stark differ-

TABLE 7.4. Assessment of OPB Positive Culture at XDOT and YDOT Based on GIS Implementation Track Decisions

GIS implementation track event	Presence of positive OPB culture		Acceptance and practice of "WIIFM"	
	XDOT	YDOT	XDOT	YDOT
Organizational location of GIS unit	+ + +	–	+ + +	–
Hiring new GIS staff	+ + +	N/A	+ + +	N/A
Attribute-data transfer to GIS	–	N/A	+ + +	N/A
Application development standards	+ +	–	+ +	–
Database construction priorities	+ + +	–	+ + +	–
Standard hardware/software platform	+ + +	–	+ + +	–
Target delivery architecture	+ + +	–	+ + +	–
Base-map scale and accuracy	N/A	–	N/A	–
Job reclassification	–(?)	N/A	–(?)	N/A

Note. The particular "positive" OPB culture/actions was present to the following degrees: ? = cannot be determined based on case evidence; – = none; + = low; + + = moderate; + + + = high; N/A = not applicable (track event did not take place or was insignificant). From Pinto and Azad (1994). Copyright 1994 by the Urban and Regional Information Systems Association. Reprinted by permission.

ences in the presence of positive OPB and practice of WIIFM: XDOT exhibited a high positive OPB culture and the practice of WIIFM; for YDOT, the opposite was true.

It is risky to offer conclusions based on only two cases. We would, however, assert that the success of GIS planning and implementation in XDOT is, in part, due to the positive OPB culture and WIIFM practice. That is, the organizational actors appear to adhere to quid pro quos regarding informal means and influences amid workgroup conflict, which tend to promote solutions that are in the interests of the whole organization. The two clear track events were the GIS unit organizational location/new staff hires and the application development standards. The attribute-data transfer mechanism did not adhere to such principles and involved dysfunctional political behavior at least as far as the whole organization was concerned.

In the YDOT case, we are tempted to interpret an almost total disregard for organization-wide concern with efficiency and effectiveness, which manifested itself through workgroup self-interest in almost all track events of GIS implementation. There was little positive OPB culture promotion and no WIIFM practice. We again attribute the "failure" and termination of the GIS project at YDOT at least partially to this lack of OPB and WIIFM practice.

The contrast between the two cases is sufficient, in our opinion, to point out the implications of the theoretical constructs and propositions as well as the practical implications of positive versus dysfunctional OPB.

Conclusions

Implementation politics is a process that few managers and professionals enjoy. We do not like having to cut deals, to negotiate the introduction of new systems, and to constantly mollify departmental heads who are suspicious of the motives behind installing GIS or any other system that threatens their power base. Nevertheless, the realities of modern organizations dictate that successful managers and professionals must learn to use the political process for accomplishing goals and implementing plans/ projects.

Our goal has been twofold, to offer a research framework as well as practical views on the nature and importance of political behavior in modern organizations. We have laid out some of the major issues in organizational political behavior, and illustrated them by means of two case studies of GIS implementation. Although two cases do not constitute adequate validation, that was not our aim. It was, rather, to provide exploratory evidence to deepen and refine the OPB concepts presented. We need to continue to refine the concept of OPB and better understand the types of behaviors that can yield positive outcomes—that is, those results that employ OPB in a positive manner. It is also important to consider this topic in light of another chapter in this book that deals with project champions. If, as we argue, championing behavior consists of using power in entrepreneurial ways, readers can immediately see the linkage between champions and effective OPB. Regardless of one's perspective on OPB, there is no doubt that it is a critical skill to master. Effective GIS implementation and use depends on it.

Chapter 8

Economic Justification
for GIS Implementation

In the years since the publication of the first edition of *Managing Geographic Information Systems*, GIS have gained more widespread acceptance and adoption. While the technology is better accepted than in the past, purchase of a GIS and the related costs of operation remain a substantial investment for most organizations, requiring significant commitment of funding, staffing, and time. A valuable technique to justify technology acquisition, including GIS, in both public and private sector alike has been the benefit–cost analysis. The basic framework of a benefit–cost analysis is to identify and then assign a price to the benefits and costs of adopting a GIS, then compare the benefits with the costs.

Organizations must perform benefit–cost analysis with care, and pay special attention to the accuracy and currency of the information they use to substantiate their analysis. This is particularly true in the public sector, which operates under somewhat different economic constraints than does the private sector because of the absence of traditional pricing mechanisms. Rather than being motivated by profit, public organizations have a mission to serve the public interest.

This chapter begins by presenting the basic elements of benefit–cost analysis, providing examples of specific benefits and costs associated with GIS implementation. The chapter continues with refinements of benefit–cost analysis, including a lesson on discounting, which is a crucial element in the GIS context because of the high front-loaded costs and the long-term benefits associated with this technology. The chapter ends with cautionary discussions regarding intangible costs and benefits associated with GIS, as well as complications in the public sector related to the lack of a pricing mechanism and the presence of a duty to the public interest.

Portions of this chapter are from Field (1994). Copyright 1994 by The McGraw-Hill Companies. Reprinted with permission of The McGraw-Hill Companies.

An Introduction to Basic Benefit–Cost Analysis

Experts in the field use the terms "benefit–cost" and "cost–benefit" more or less interchangeably to describe the technique. Using "benefit–cost" has its advantages, however, the most obvious of which is that this word order implies that benefits must outweigh costs in order to justify a particular course of action. Moreover, some scholars (e.g., Zerbe & Dively, 1994) argue that the term "benefit–cost" implies a richer analysis than the alternative phrase. "Benefit–cost," therefore, is the phrase that we use in this chapter.

The end of the 20th century brought with it a growing emphasis on economic efficiency within organizations in the private and the public sectors alike. "Doing more with less," "downsizing," and "rightsizing" all became part of everyday language as euphemisms for budget cutting and layoffs. Moreover, corporate scandals at the dawn of the 21st century have focused new attention on corporate accountability. In response to calls for increased efficiency and accountability, organizations across the board must provide reliable and defensible justifications for every purchase or new initiative they undertake. Benefit–cost analysis is the first-line procedure to assure that organizational GIS initiatives are justifiable; indeed, many organizations require a benefit–cost analysis as a prerequisite to adopting any new technology (Huxhold & Levinsohn, 1995).

GIS (and other information technologies) have never been a better buy than they are today. Declining prices, especially of hardware, have accompanied an explosion in the computing power of GIS that have become remarkably user-friendly in recent years. The potential of GIS to improve the overall efficiency and productivity in organizations whose missions rely on geographically referenced data has never been greater. Ironically, the current economic and corporate environment also means that the need to justify institutional action, including the purchase and implementation of GIS, also has probably never been greater.

The use of benefit–cost analysis to justify the adoption and implementation of GIS is well established in the GIS literature (e.g., Aronoff, 1989; Dickinson & Calkins, 1998, 1990; Grimshaw, 1994; Huxhold, 1991; Huxhold & Levinsohn, 1995; Obermeyer, 1999; Smith & Tomlinson, 1992; Wilcox, 1990). Traditional benefit–cost analysis as an economic exercise begins with an organization's identifying and listing the costs associated with implementing a GIS. These costs include expenditures on computer hardware and software, the transformation of paper maps and data into digital format, and hiring additional GIS specialists or training existing staff members to use the technology effectively. Equally important to the analysis, the organization must identify and list the expected benefits of implementing the technology. These benefits include improved efficiency and effectiveness. We describe typical GIS benefits and costs early in this chapter.

The next step in the analysis requires that the organization assign economic value (price) to both the costs and the benefits, sum each of them, then compare the results mathematically, normally using dividing the value of the benefits by the value of the costs (benefits/cost = n). If the value of the benefits is equal or greater than 1, the organization has successfully justified the investment in the technology. Benefit–cost analysis for high-cost expenditures with long-term benefits typically covers a multiple-year period. Payback period and discounting are normally used as refinements in the implementation of multiyear projects. This is particularly true for the organization attempting to justify the implementation of a GIS because of the high early costs and enduring benefits of the technology.

A note: Some scholars have criticized benefit–cost analysis as a "dogmatic approach that knows the price of everything and the value of nothing" (Zerbe & Dively, 1994: xv). Not surprisingly, the most thoughtful justifications approach the benefit–cost analysis as an art that recognizes the importance of ethics and values as well as the need to consider more interesting and complex questions of economic theory (Gillroy, 1992; Zerbe & Dively, 1994). Huxhold and Levinsohn (1995), for example, recommend an examination of the financial, technical, and institutional feasibility as an alternative to benefit–cost analysis, while Grimshaw (1994) suggests a value-added approach to justify a GIS.

Benefits and Costs in the Analysis

A typical benefit–cost analysis contains several elements. As previously noted, the most rudimentary element is the identification and assignment of a numerical economic value to the benefits and costs associated with an initiative. Costs must include any expenditure that the organization incurs as a result of implementing the project: purchase of any hardware (including computers and all peripherals), software and related supplies, the cost of hiring any additional staff or the training of existing staff, along with the cost of transforming maps and data into digital format.

Costs of these types are classified as "tangible" costs. Tangible costs are defined as being readily quantifiable, primarily because they represent costs of products that are bought and sold in the free market, even when the organization implementing the technology is a public agency.

Along with costs, many of the benefits that the organization will experience are also tangible, and therefore relatively straightforward to quantify. For example, if the organization expects to be able to reduce its workforce because of the increased efficiencies that implementation of GIS technology promises, the organization will be able to look up the cost of the salary (or wages) plus benefits of staff members whose services may no longer be required. Similarly, if the GIS will enable an organization to produce more detailed or more diverse information and information products as a result of implementing a GIS, the organization should also be able to improve its overall effectiveness. Thus, the first cut at benefit–cost analysis is the easiest: quantifying the tangible costs and benefits. To the extent possible, the organization must use objective data in the analysis, and verify data and data sources (Dahlgren & Gotthard, 1994). As one might expect, the analysis usually becomes much more complicated.

Benefits

There are three major categories of benefits that a benefit–cost analysis for GIS should examine: (1) cost reduction, (2) cost avoidance, and (3) increased revenue (Huxhold, 1991: 244). Aronoff (1989: 260–261) identifies five: (1) increased efficiency, (2) new nonmarketable services, (3) new marketable services, (4) better decisions, and (5) intangible benefits. (See Table 8.1.) Aronoff's ideas of increased efficiency and new marketable services correspond to Huxhold's notions of cost reduction and increased revenue, respectively. It is important to recognize, however, that price reductions made possible by the lower costs associated with GIS implementation may actually stimulate demand for some geographic information products. This can result in increased revenues overall because of increased volume of sales (Rhind, 1996). We will say more about the sale of geographic information products later in this chapter.

Huxhold (1991: 244) defines cost reduction as "the decrease in operating expenses of the organization, primarily caused by a savings in time by operating personnel performing their tasks more efficiently." Cost reductions generally accrue because of the improved productivity of staff members responsible for the tasks performed using the GIS.

TABLE 8.1. Benefits from GIS Adoption

Huxhold	Aronoff
1. Cost reduction	1. Increased efficiency
2. Cost avoidance	2. New nonmarketable services
3. Increased revenue	3. New marketable services
	4. Better decisions
	5. Intangible benefits

Note. Data from Huxhold (1991) and Aronoff (1989).

Cost avoidance is the "prevention of rising costs in the future caused by projected increases in workload" per staff member (Huxhold, 1991: 246). This benefit is consistent with, and more or less an extension of, the first benefit, suggesting that once a GIS becomes part of an organization's equipment, it may help to optimize the performance of a variety of both current and future tasks. This improvement in performance may make it unnecessary to hire new employees, or at least to postpone such hires by making the best use of existing employees (246).

As an example of cost reduction or avoidance (Aronoff's improved efficiency), consider the benefits of using a GIS to answer routine questions about parcel ownership in a county tax assessor's office. Typically, answering questions as simple as "Who owns the property at 100 North Main Street?" can involve looking up cross-referenced information in two or more hard-copy record books. This can take several minutes. In contrast, a search employing GIS technology can be provide nearly instantaneous results, saving many hours of staff time over the course of a year. Moreover, requests for maps of the property at 100 North Main Street can be handled simply, effectively, and quickly. Cities such as Milwaukee, Wisconsin (*www.gis.ci.mil.wi.us/Map_Milwaukee/map_milwaukee.htm*), and Minneapolis, Minnesota (*www.metrogis.org/*), have online GIS, which provide round-the-clock access for the general public to a wide variety of geographic information products, along with the answers to many routine questions from taxpayers and citizens. By the way, William Huxhold (oft-quoted in this book) was instrumental in the adoption and implementation of Milwaukee's GIS.

Finally, Huxhold (1991: 245) suggests that "a GIS can increase revenues . . . by selling data and maps, increasing property tax collections, and improving the quality of data used to apply for state and federal grants." The rationalization of tasks that the GIS makes possible does indeed bode well for the increase in tax collections and the improvement in data quality. However, Dansby (1991) suggests that there may be legal impediments to the sale of such products in the public sector, depending on state and local regulations regarding copyrights and freedom of information considerations. Any state or local organization that wishes to sell GIS products needs to discuss the issue with its legal department.

In general, the U.S. federal government is not allowed to sell geographic information for profit because the people of the United States, rather than the government, own the data. This is not true in other countries. The United Kingdom, for example, sells maps and related products at market rates. This difference has consequences. In the United States, maps prepared by the federal government are available at low cost, an amount roughly equivalent to the cost of reproduction and handling. In some instances, maps are free. Purchasers may use the maps as end-products, or they are free to add value and sell the new map products at market rates.

New nonmarketable services are "useful products and services that were previously unavailable" and will be used within an organization, perhaps to make the organization run more smoothly, or to provide additional information that will improve decision making (Aronoff, 1989: 260). Aronoff points out that organizations will be able to anticipate some of these benefits of GIS in early discussions about the purchase of the technology. Other benefits, however, will not become apparent until after the GIS is up and running. Therefore, it will be difficult to assess the value of nonmarket services and include accurate figures in the GIS benefit–cost analysis. The inability to place a definitive value on a nonmarketable service should not prevent an organization from at least noting the service and providing a qualitative description of its benefits.

As noted, because of the cartographic capabilities embedded in today's GIS, implementation of the technology will make possible the generation on demand and subsequent sale of new geographic information products. These new products are the result of the inherent ability of GIS to extract and combine data in a variety of combinations and permutations, essentially enabling its implementors to produce customized maps and geographic information products on demand. For example, a city government with a comprehensive, large-scale GIS with current, accurate information can quickly produce a map of vacant downtown retail space for an individual wishing to open a bookstore, along with a table identifying the owners of the properties identified.

Aronoff (1989) also suggests that the adoption of a GIS will produce "better decisions" (261). This will occur, he believes, because "more accurate information and faster and more flexible analysis capabilities can improve the decision-making process itself" (261). Again, determining the economic value of these "better decisions" that may occur in the future is difficult at best. Moreover, Aronoff's prediction of "better decisions" resulting from GIS adoption is optimistic. The large body of literature on organizational decision making takes a more realistic view, essentially conceding that most decisions are made on the basis of incomplete information, usually biased toward information that is familiar or comes from familiar sources (see, e.g., Cyert & March, 1963; Douglas, 1986; Downs, 1967a; Simon, 1945/1976). In some cases, organizations deliberately limit their searches for information because of time and/or financial constraints. In other instances, organizations may be unaware of additional relevant information (seeking and using information have costs). A GIS alone cannot eliminate these institutional limitations.

Costs

There are several costs associated with implementing a GIS: hardware, software, technical support, transforming maps and data into digital format, staffing the GIS, and intangible costs. (See Table 8.2.) The most obvious, perhaps, is the cost of new hardware. While most organizations have personal computers these days, existing equipment may not be up to the challenge of running GIS software, even though more GIS run on PCs. High-end GIS software continues to use workstation environments. Even GIS software that runs on PCs generally has requirements for speed and data storage capacity that exceed the general purpose computers in place at most organizations. Furthermore, given the huge storage needs associated with GIS, it is best to have at least one dedicated computer.

Additional hardware is also warranted with the installation of GIS. Scanners, digitizing tablets, large-format printers, and global positioning systems are some of the hardware that may be useful. The type of equipment needed, and its cost, will depend in part on the tasks for which the GIS will be used and the software that the organization chooses.

It goes without saying that GIS software is a key part of the implementation. In choosing GIS software, an organization is well advised to begin with a clear understanding of the tasks the GIS will perform to make certain that the software is up to the challenge. Keep in mind that upgrades to the software will most likely also be forthcoming, and are usually part of a package deal. Moreover, these costs will add to the annual cost of maintaining the GIS. In addition to meeting with software vendors, orga-

---

TABLE 8.2. Costs of GIS Adoption

1. New hardware (computers, scanners, digitizing tablets, printers, etc.)
2. Software
3. Technical support
4. Transforming maps and data into digital form
5. Hiring GIS-proficient employees or GIS training for existing employees
6. Intangible costs

---

nizations should also seek the advice of organizations similar to itself that have already adopted GIS in order to learn what they can from their experiences.

Closely related to software is the cost of technical support from the software vendor. Like software upgrades, this may be provided as part of a full-service package deal. It will be necessary to know what is (and is not) included in the package. Like software upgrades, technical support will also normally add to the annual cost of maintaining the GIS.

Data are another critical element of the GIS. Without data, the GIS cannot exist. While there is a wide variety of high-quality digital data available, some of it at low (or even no) cost, it is virtually guaranteed that every organization will need to collect new data that are specific to the needs of the organization's GIS, and/or digitize existing analog data. The cost of collecting and digitizing data will depend on several factors, including the level of accuracy and precision that the GIS requires. Of course, the more accurate and precise the data needed, the higher the cost will be. The organization must know at the outset of its efforts to implement a GIS what its data needs are. For example, a utility provider, which has responsibility for providing uninterrupted service to its customers and "call-before-you-dig" programs, must know to the centimeter where its lines are buried. On the other hand, a nongovernmental organization that is using GIS to monitor local planning board decisions on specific parcels does not need to know the location of map objects to the same small margin of error required by utilities. Knowing and understanding the accuracy and precision requirements needed for your specific GIS implementation is exceedingly important. On the one hand, this is important because of the need to collect data of sufficient quality to support the GIS's functions. On the other, it is important because there is a direct correlation between data quality and price.

Providing staffing for the GIS is an ongoing problem. In Chapter 12 we discuss this issue in greater detail. However, at this point it is important to note that implementing a GIS is likely to require enhancing GIS expertise within the organization. This is normally accomplished in one of two ways. The first way is to hire one or more GIS specialist (perhaps a GIS manager) from outside the organization. The other method is to train existing members of the organization. Murray (2001) advises that training is needed to maximize the investment in any information technology. Like software upgrades or technical support, software training is normally available from the GIS software vendor; it may even be possible to negotiate training as part of a longer term package.

Just as there may be intangible benefits associated with GIS implementation, there may also be intangible costs. For example, there may be disruptions in service associated with switching from manual to automated transactions that the GIS makes possible, or in transforming analog data into digital format. Individual members of the organization may have different reactions to the transition to GIS. While some are likely to be enthusiastic, others may be uncertain about their changing role, or suspicious of the project. Another possible effect of the GIS implementation is that there may be shifts in the individual roles and assignments of various members of the staff, especially if the organization hires new GIS professionals. Murray (2001) suggests that including the IT department, related units, senior management, and other relevant units in the GIS implementation process may help to minimize these intangible costs. While open discussions at all stages of the implementation may minimize these (and other) intangible costs, the organization must be prepared for glitches in the process.

Once the organization catalogs these basic benefits and costs, it can go on to further refinements of the benefit–cost analysis. These are described in the remainder of this chapter.

Refinements of Basic Benefit–Cost Analysis

There are several variations on benefit–cost analysis; one is cost-effectiveness analysis (Layard & Gleister, 1994). Cost effectiveness analysis provides a comparison of the costs of providing a specific outcome, or performing a specific task, using different means. In adding this step to the benefit–cost analysis, the organization compares the costs of alternative means of performing the same task—for example, the cost of providing information on property ownership both with and without a GIS. Implicitly, adding this step forces the organization to demonstrate not just that the benefits of its initiative outweigh the costs, but that a specific strategy for performing a specific task is more cost-effective than other strategies (Layard & Gleister, 1994: 21).

A second variation is the calculation of the "payback period" (Huxhold, 1991: 249). Calculating the payback period begins by dividing the total cost of implementing a GIS by the estimated annual value of the benefits of using the system. The resulting figure tells how many years it takes to accumulate enough benefits to pay for the cost of the system (Huxhold, 1991: 249). The benefits may include any or all of the benefits described earlier in this chapter. Not surprisingly, this calculation is fraught with the same difficulties apparent in typical benefit–cost analyses since some of the costs and benefits are subject to speculation or at least debate.

Grimshaw (1994: 121) endorses a third variation, the value added approach. This approach emphasizes the new things technology enables the organization to do, and what these new capabilities add to the value or worth of the organization. This strategy echoes Aronoff's new marketable and nonmarketable services.

Murray (2001) recommends developing alternative implementation scenarios in order to encourage discussion. These alternatives start with a baseline level of implementation, but describe medium and high-end implementations as well. With each alternative that exceeds the baseline, the organization should describe (and evaluate to the extent possible) what this additional service would add to the organization's mission. Furthermore, Murray recommends a "gap analysis" to determine the difference between the current state and the ultimate goal of the IT investment, along with a plan to bridge the gap. The implementation plan should include a time frame, along with information on elements that decision makers considered but rejected.

Several problems arise in performing cost–benefit analysis, some of which apply across the board, others of which are unique to the public sector. There are several refinements of the process to address these difficulties.

Time and Discounting

One problem that arises in performing benefit–cost analysis is caused by the effects of time and economic inflation (Field, 1994; Layard & Glaister, 1994; Little & Mirrlees, 1994; Smith & Tomlinson, 1992; Stiglitz, 1994).

Even when the rate of inflation is low, over time the cumulative effects of inflation erode the economic value of the costs and benefits of any activity. Moreover, people perceive immediate benefits as having greater value than benefits far off in the future. As Zerbe and Dively (1994: 43) put it, "a benefit received today is worth more than one in the future." Similarly, a cost that occurs far in the future has far less significance than a cost today (Field, 1994). In order to provide a realistic assessment of costs and benefits, organizations must take the effects of time into account and adjust their benefit–cost calculations accordingly.

A refinement designed to address this problem is discounting (Field, 1994: 119–123; Smith & Tomlinson, 1992: 255–256). The idea behind discounting is to deflate the costs and benefits in order to remove the effects of inflation. In short, discounting provides a mechanism to address the old saw that money doesn't go as far as it used to. Discounting is needed to provide an accurate assessment of the value of implementing a GIS because of (1) the multiyear life expectancy of a GIS, and (2) the resulting fact that GIS costs and benefits are also spread over multiple years.

Still, discounting is not a simple matter, particularly with GIS, which have their largest outlays early in the life of the project, then experience declining costs, but whose benefits last long into the future. Front-end costs include the purchase of hardware and software and either hiring new staff or paying to educate existing staff. In addition, an organization can expect high start-up costs arising from the need to put analog maps into digital form. These start-up costs are likely to seem insurmountable for many small- and medium-sized cities. The perception of insurmountable costs is likely to be compounded by the recognition that the local government will not begin to realize the benefits of a GIS (which, as noted, are enduring) for several years.

Discounting applies to both costs *and* benefits. Its primary purpose is to aggregate a series of costs and/or benefits that occur over the life of a project (Field, 1994). The formula for discounting includes three elements: present (or future) value, the length of time appropriate for the project, and an appropriate discount rate (Field, 1994).

$$\text{Present value} = \frac{\text{Future value}}{(1 + \text{discount rate})^{10}}$$

To calculate the present value of $1,000 10 years in the future with bank interest rates at 5%, the formula is as follows (Field, 1994):

$$\text{Present value} = \frac{\$1000}{(1 + .05)^{10}} = \$613.90$$

Discounting over a multiyear project like a GIS is more complicated because of the high front-end, but then declining costs and the low front-end, then rising and finally steady benefits of implementing the technology. Multiyear projects are handled as in the following example, a hypothetical GIS implementation. Assume that the costs and benefits for the first 7 years of the project are as shown below, and that the discount rate is 6%:

Year	1	2	3	4	5	6	7
Costs	100,000	70,000	50,000	25,000	25,000	25,000	25,000
Benefits	0	25,000	70,000	70,000	70,000	70,000	70,000

Calculate the present values of costs and benefits using the following formula (Field, 1994):

$$PV_{cost} = \$100,000 + \frac{70,000}{1+.06} + \frac{50,000}{(1+.06)^2} + \frac{25,000}{(1+.06)^3} + \frac{25,000}{(1+.06)^4} + \frac{25,000}{(1+.06)^5} + \frac{25,000}{(1+.06)^6}$$

$$PV_{cost} = \$0 + \frac{25,000}{1+.06} + \frac{50,000}{(1+.06)^2} + \frac{70,000}{(1+.06)^3} + \frac{70,000}{(1+.06)^4} + \frac{70,000}{(1+.06)^5} + \frac{70,000}{(1+.06)^6}$$

Choosing a discount rate is not a simple matter. First, there is the issue of real versus nominal interest rates. Nominal interest rates are the actual interest rates available in the market. In order to know the real interest rates, it is necessary to adjust these nominal figures for inflation. For example, if the nominal interest rate is 8%, but the average rate of inflation over the period in question is 3%, then the real interest rate is 5% (Field, 1994). In all instances, managers must always consistently use either real costs and real discount rates, or nominal costs and rates (Field, 1994).

The large number of interest rates in use in the world of modern finances complicates the process of discounting. A review of the business/finance section of any reputable newspaper shows a large variety of interest rates from which to choose: rates on normal savings accounts, certificates of deposit, bank loans, and government bonds, to name just a few. There are two views on this issue. The first view suggests that the discount rate should reflect the way people think about time and money.

Economists refer to this as "the rate of time preference." For example, most people would prefer receiving $1 today, rather than waiting 10 years to receive that same amount. This is a positive rate of time preference. Those who support this view would use the average interest rate on a bank savings account as their discount rate (Field, 1994).

The second approach to choosing a discount rate is based on the notion of investment productivity. In this view, people anticipate that the value of future returns will offset the cost of investment today. In the public sector, this means that expenditures used for long-term projects should yield rates of return to society that are similar to what the same expenditures could have earned in the private sector (Field, 1994). Using this reasoning, an organization should use a discount rate that reflects the rate banks charge their investment borrowers; these rates are typically higher than savings account rates (Field, 1994).

The debate about discount rates ultimately leaves it up to the manager to choose—and justify—an appropriate discount rate. One resolution is to perform a sensitivity analysis by repeating the discounting of benefits and costs using two or more different interest rates.

It is not difficult to grasp the impediment that discounting imposes on a benefit–cost analysis for GIS. The high start-up costs of GIS will seem even higher than they are in light of the positive rate of time preference. On the other hand, the benefits of GIS will seem smaller after discounting. If one carries out the calculations on the hypothetical 7-year GIS implementation example provided above, it will take the entire period for benefits to begin to outweigh costs. A real-life GIS may take even longer to reach the break-even point.

But it is important to remember that the benefits of GIS are enduring. Once an organization has paid the high front-end costs, particularly those associated with higher staffing costs and digitizing, it will reap the benefits of the technology year in and year out. It is therefore necessary to emphasize the enduring nature of the benefits of GIS. This can be accomplished by carrying out the analysis for as many years as required to achieve a favorable benefit–cost ratio. In addition, however, the manager should also make it clear that transforming data into digital format is a one-time-only expense. It is also important to review the benefit–cost analysis at the completion of each major phase of the project life cycle to make certain that the implementation continues to produce a favorable outcome (Dahlgren & Gotthard, 1994). Finally, the manager should make clear that the investment in GIS will endure for generations to come.

Stakeholders

Within the context of any organization's mission, there are a variety of individuals and/or groups who have an interest (or a "stake") in what the organization does and the strategies it employs (Layard & Glaister, 1994; Sen, 1994; Zerbe & Dively, 1994). An organization's customers or clients are a crucial component in its survival (Obermeyer, 1990a; Weber, 1946). The costs and benefits of the actions of an organization may not be identical for all the different individuals or groups with a stake in the organization's actions.

For example, a company whose mission is to produce road maps includes among its stakeholders individuals and groups with varying needs for map detail. The average user who has found the company's maps to be excellent navigation aids is unlikely to be impressed if the company makes a decision to provide more detailed maps if that additional detail comes at a higher price—even if the price is only slightly higher. If the company has a competitor that produces a map comparable to the original map at a price lower than the "new and improved" (and more expensive) version, the company may, in fact, lose market share and perhaps suffer declining revenues as a result of its decision to offer greater detail at a higher price.

The stakeholder problem is even more complex in the public sector, where levels of income among end-users vary greatly (Layard & Glaister, 1994). For example, a professional nature photographer who can afford to hire a native guide to lead her to the lair of an endangered animal (and may also be able to deduct the cost of the guide as a legitimate business expense for tax purposes, too) has no real need for a detailed, large-scale map of the area. On the other hand, a PhD candidate trying to study that same animal would probably find such a map to be essential. Thus, trying to account for the costs and benefits to all stakeholders can become a complicated task.

Certainly, a manager cannot afford to ignore the organization's various stakeholders. On the one hand, to ignore stakeholders is to risk alienating existing and potential customers and clients. On the other hand, the organization may miss an opportunity to report higher benefits arising from its ability to enhance the satisfaction level of existing stakeholders, or by increasing the actual number of stakeholders reported in its benefit–cost analysis. The flexibility of GIS may, in fact, make it possible for both private and public organizations to increase their product lines and fill new market niches at relatively small additional costs, and, as a result, to increase their customer and client bases by appealing to a wider audience. For example, Rand McNally produces road atlases of the United States and Canada with varying levels of detail and typeface. Its "regular" atlas features a photograph of a young couple in a convertible sports car. The large-type version has a cover photograph of an older couple (clearly senior citizens, though also clearly active). Although one of the authors purchased the latter for her children's use on a road trip to Texas, it was only later that she noticed the marketing to these two different customer bases.

It is up to the manager to estimate the expected value of these potential benefits and include them in the analysis. For example, the director of a local planning agency can build a case for a GIS by first identifying, then estimating the value of the GIS to the local government itself following Huxhold's categories, for example. However, the availability of a large-scale, comprehensive GIS will also benefit local utilities, developers, and private businesses by making accessible high-quality "official" geographic information products that these groups can then use to inform their own decisions and to help in their day-to-day operations.

Not surprisingly, some local governments have exploited the relationship with their stakeholders by working cooperatively with groups such as local utilities and business leaders to build and implement their GIS. For example, the Cincinnati Area GIS (Cincinnati, Ohio) is a joint venture of the city and county governments, the telephone company, the local power and water companies, and local industry (which includes Proctor & Gamble) (Obermeyer, 1995). Working with stakeholders has the added advantage of sharing costs among the participants and improving the level of benefits as a result of the specific functional expertise—and data—that each participant brings to the project.

Uncertainty and Risk

Time also influences the level of risk and uncertainty among the benefits and costs of an organization's initiatives. Since humans do not possess perfect knowledge about the present, it is unrealistic to expect them to foresee the future. Zerbe and Dively (1994) identify two types of uncertainty: (1) uncertainty caused by the unpredictability of future events; and (2) uncertainty caused by limitations on the precision of data. Both types of uncertainty are relevant to GIS benefit–cost analysis, particularly in the past. Throughout much of the time since GIS has become commercialized, there has been a great deal of uncertainty about both the costs and the benefits of the technology. For example, on the one hand, lack of experience in the early days of commercial marketing of GIS meant that many organizations underestimated the long-term costs of the implementation, such as digitization costs, consultation fees, and training expenses that often far surpassed initial estimates. On the other hand, today's turnkey GIS products (e.g., "Maptitude" and "Arcview") enable GIS adopters to know with greater certainty the cost of the basic package. However, there remains a great deal of uncertainty associated with other critical elements of GIS start-up, namely, hiring and/or training staff, digitizing maps, and gathering and entering data to customize the GIS.

In evaluating the wisdom of purchasing a GIS, both the benefits and the costs of implementation may be difficult to assess because of the uncertainty surrounding them. It is well known and generally accepted that the costs of implementing a GIS extend beyond the purchase of hardware and software. For example, assembling and maintaining data and training staff are two areas that will require expenditures after the initial purchase of the GIS. The exact dollar amount of these additional costs is usually difficult to know ahead of time. However, as Smith and Tomlinson (1992: 254) optimistically note, "The costs [associated with implementing a GIS] are loaded heavily in the early period whereas the benefits increase . . . and then remain constant." This assumes a stable organization and external environment. The wise manager will prepare for unexpected contingencies throughout the life of the system.

The uncertainties surrounding the calculation of benefits and costs of implementing a GIS have been the subject of discussion by several authors (e.g., Aronoff, 1989; Huxhold, 1991). There are several approaches to handling uncertainty in benefit–cost analyses. The first is to ignore it, which is appropriate if the uncertainty is minor, or where the analysis is intended to be only a rough estimate. It may also be possible to reduce uncertainty by gathering additional information, and the organization should make every effort to do so. The project manager should also talk with other similar organizations that have implemented GIS in order to add to their knowledge base. Finally, the organization can recognize uncertainty and factor it into the benefit–cost analysis explicitly (Zerbe & Dively, 1994: 371).

Selling Data

The sale of geographic information products is often suggested as a "benefit" to be included in a benefit–cost analysis. Properly managed, these benefits can indeed be significant. For example, David Rhind (1992) reports that Great Britain's Ordnance Survey generates $100 million in annual revenues through the sale of geographic information products. Ownership of the copyright to data sets is a prerequisite to having the right to make such sales. In most countries, the national government holds the copyright to all such data sets they develop; the U. S. federal government is an exception to this rule (however, U. S. cities and states can copyright data).

There may be difficulties in establishing prices for geographic information products, but an organization can compare its geographic information products with similar products offered for sale by the private sector in order to establish a basic price list. Once products are officially offered for sale, the organization can adjust the price to try to achieve its desired sales and revenue goals.

Sale of data and other geographic information products may expose an organization to liability risks arising from negative outcomes associated with unintended uses or deliberate abuse of the products. The wise manager will consult with the organization's legal department to resolve these issues.

In short, organizations contemplating the sale of data as a benefit of their GIS should be aware of the pitfalls as well as the benefits. The potential rewards certainly warrant the sale of geographic information products, if the organization is permitted to do so.

Whether or not it is legal to sell GIS products created in the public sector depends on the applicability of two conflicting laws in each specific case: freedom of information (open records) and copyright protection (Dansby, 1991: 100). State and local governments have the same copyright protection available to individuals and other entities; the federal government, however, does not. Dansby suggests that, in general, GIS databases developed by state and local governments could be considered as original works of authorship, which therefore would be protected by copyright. In cases where geographic databases contain material that originates with third-party authors, the third-parties (not the government agency) would own the copyright to the data. The ownership of a copyright is a prerequisite for selling products for cost-recovery purposes. Therefore, governments that own the copyright to the information in their GIS databases have met the first condition for selling the information.

In potential conflict with copyright laws, however, open records laws frequently include clauses that prohibit the sale of the records for any more than a nominal fee (e.g., the cost of reproduction) (Dansby, 1991: 101). Dansby argues that the validity of using copyright protection as a rationale to charge fees for GIS products or other government data "depends on the wording of each state's open records laws. . . . [N]othing can prevent cost recovery to help finance GIS development in those states whose open records laws are narrowly drawn to allow inspection only. Other states would require that exceptions to the open records laws be passed or that the laws be amended" (101). In short, the potential benefits of selling GIS products are not available to every jurisdiction that employs a GIS, and therefore should not be included automatically in every benefit–cost analysis.

Moreover, there is an ethical argument against selling government data at a profit, based on concerns about charging the public twice for the same information. In most cases, in the United States, taxes fund federal government activities, including the development of databases and GIS. These taxes come from people living in the jurisdiction, and may come in the form of income taxes, sales taxes, property taxes, or fees for use. The argument that it is best to charge merely for the cost of reproduction of information rests on the acknowledgment that the citizens have already paid to gather and develop the information. Any profit taking would appear to be double charging, and may alienate the citizenry.

On the other hand, making available at a nominal cost information that a purchaser may use in a secondary profit-making activity raises a different set of ethical problems. Is this an appropriate use of data collected at public expense? Should the government subsidize individual profit-making activity? But isn't the purchaser who adds value and resells government data also a member of the public? Since these entrepreneurs have contributed to the development of the data through their tax payments, should they be deprived of the use of these maps and related data for profit-making activity? Won't these entrepreneurs pay taxes on the income they make from these activities, and otherwise contribute to the economic well-being of the country? Royalties, as opposed to outright purchase, have been suggested as a means to address this concern.

Additional concerns about selling data arise regarding liability for outcomes of unintended uses or deliberate misuses or abuses of the data. Once data and information find their way into the hands of purchasers, the developers of the maps and data no longer have control. Unintended and inappropriate uses, as well as deliberate abuse of data and information, may result in negative consequences. Who is liable? The purchaser only? Or is the government agency that sells the data also liable? With potentially large cash settlements at risk, this is not a trivial matter.

The problems attending the potential benefit of selling data suggest that there is great uncertainty associated with including this "benefit" in any benefit–cost analysis. Many details of such sales remain unresolved. The passage of time and the accumulation of experience will lead to the drafting of appropriate legislation and the development of case law to address these problems. Until then, it is wise to use extreme caution in including sale of data and information as a "benefit."

Externalities and Spillovers

Externalities and spillover effects are mirror-image problems that may arise in developing benefit–cost analyses. Externalities arise when a firm shifts its costs outside the organization, usually by ignoring a problem (Papageorgiou, 1978). Externalities are particularly troublesome for public institutions, which are limited in their ability to externalize, yet frequently are the only hope for cleaning up problems created when private organizations externalize their costs. For example, in the United States, the federal government has assumed responsibility for cleaning up toxic waste dumps created by the private sector. It is true that the government could ignore the problem, but this strategy could lead to problems cropping up elsewhere—for example, in the overall health of people living near the sites. These are ramifications that private firms can—and often do—ignore.

Spillover effects, or positive externalities, are the benefits that an organization enjoys because the activities of another organization extend beyond its jurisdictional boundary (Faulhaber, 1975). Private firms often enjoy the spillover effects created by public expenditures (e.g., transportation networks, sewer and water projects), just as some public agencies may benefit from the activities of private firms or other jurisdictions. For example, a GIS software developer that includes government census data with its software is able to add value to its product and thus receives tangible economic (spillover) benefits from the government's data-gathering and dissemination activities.

Handling externalities and spillover benefits in the benefit–cost analysis is a matter that merits mention. In the case of governments that are performing a benefit–cost analysis as a prelude to their implementation of a GIS, Smith and Tomlinson (1992: 250) recommend incorporating "all benefits . . . in the analysis whether or not they accrue to the potential GIS purchaser or to the departments that will use the information products." Among the nongovernment groups that may realistically expect to benefit from the implementation of a GIS are taxpayers, private firms, and special service districts (Smith & Tomlinson, 1992: 250).

How does an organization handle these externalities and spillovers? First, it is necessary to identify them. Perhaps the most significant externality of a GIS is the potential loss of privacy associated with the ability of GIS to disaggregate data. Large public data sets based on national censuses are most likely to raise privacy concerns. However, some private firms have collected large databases that may also threaten the privacy of individuals. It is extremely difficult to place a value of this potential loss of privacy to an individual. Is it $1 per person? $10? More? In this instance, managers are left to make their own assessment.

Spillover effects of a GIS, as Smith and Tomlinson (1992) note, may accrue to taxpayers, private firms, and special service districts as they reap the benefits of readily accessible maps, data, and other geographic information products made possible because of the implementation of a GIS. Spillovers, while still problematic, are a bit easier—and obviously more pleasant—to handle. For example, the county assessor might anticipate shorter transaction times for fulfilling requests for basic information, such as a plat map. In order to assess the value of these time savings to customers and clients, one should multiply the average number of annual transactions by the economic value of the anticipated time savings per transaction, which in turn is based on the average hourly wage figure for the region. Given the range of beneficiaries of spillover effects, there is great value in paying careful attention to assigning benefits to spillovers. Governments, in particular, since they have a broad (and in some cases nearly universal) set of stakeholders, can bolster their anticipated "benefits" by considering spillovers.

Handling Intangible Benefits and Costs

Many of the benefits and costs that contribute to the development of a benefit–cost analysis are intangible. For example, how can one place a numerical economic value on increased reliability or institutional confusion? Aronoff (1989) agrees that adopting a GIS may bring intangible benefits to an organization. Smith and Tomlinson (1992: 249) define intangibles as "not as much a separate category of benefits as they are a class of benefits that is more difficult to quantify." These benefits might include such things as better internal communication in the organization, improved morale, and a better public image (261). Obviously, placing a precise dollar (or Deutschmark or franc or pound) value on these intangible benefits is not possible. Still, it is necessary to give an estimate. Organizations may begin by describing these potential benefits and costs in text accompanying the benefit–cost analysis.

Assigning an economic value to intangible benefits is part of the "art" of the benefit–cost analysis. Assigning such value may be accomplished by using surrogates. For example, improved morale may result in reduced staff turnover, which in turn results in lower costs for personnel searches and training. These items are easier to value than morale.

Organizations may experience negative changes as they implement GIS (Grimshaw, 1994; Huxhold & Levinsohn, 1995), resulting in additional intangible costs. For example, an organization may find as it introduces GIS that those who are most knowledgeable become more important to the organization; conversely, those who are slow to accommodate to the technology may find themselves losing ground and eventually their jobs. The overall result may be institutional confusion, which may in turn temporarily cause a drop in productivity. While a manager might find it impossible to place a precise economic value on institutional disarray, assigning an economic value to time lost to the disruption of the social order of the organization is easier to do.

Similarly, organizations may find that their foray into the world of GIS may give them increased visibility and an enhanced reputation. For example, the U.S. Geological Survey notes the value to society of improved decisions made possible by its many mapping products (Bernknopf et al., 1993). Given the zeal of the calls for "downsizing" the public sector, solidifying this relationship makes sound organizational sense, especially in light of the importance of the relationship between organizations and their client groups (Obermeyer, 1990; Weber, 1946).

Again, in assigning an economic value to these intangible benefits, organizations need to be creative. For example, public organizations that make available low-cost or even free geographic information products to citizens might place a value on the goodwill they generate through these actions by calculating the aggregate cost savings that their customers or clients received by using the organization's products rather than more expensive commercial alternatives.

Assigning value to intangible benefits and costs can be difficult. In the case of benefits, it is extremely important to do so, in order to accumulate benefits to offset costs as part of the analysis. In the case of costs, it is necessary to do so in order to achieve fair and honest results. As noted, this part of the analysis is as much art as it is science. Still, through careful thought, an organization can assign plausible and defensible values to these intangibles.

Conclusions

The purpose of this discussion of benefit–cost analysis is to briefly describe the method and to point out its advantages and disadvantages. Its primary advantage is that it provides an economic rationale for an organization's decision to adopt a GIS. This can be very persuasive. The disadvantage of using benefit–cost analysis is that it is very difficult to assess or quantify many of the benefits and costs that are needed to perform the calculation. Organizations that wish to do such an analysis may find it helpful to follow the example of other organizations that have undertaken this task previously and, obviously, to be cautious in making their own assessments.

There remain other tidbits of advice worthy of passing along. Murray (2001) acknowledges that organizations must always seek to align funding for IT with the business goals of the organization. Furthermore, he makes an important point when he suggests that "reaching the right level of IT funding is not an exercise in expense reduction, but in moving to higher levels of IT performance" (29). This advice is relevant to GIS implementation, since this technology is inherently designed to improve the way the organization does business. Dahlgren and Gotthard (1994) also suggest that preparers of benefit–cost analyses must be certain that their methods and results are comprehensible to nontechnical professionals who must review their research.

More than a decade after the original publication of *Managing Geographic Information Systems*, the technology has become far more familiar, and there may be an inclination for some organizations to believe that they must jump on the GIS bandwagon. While implementing GIS may very well be the right decision for an organization, each organization must still justify its decision to make the move to GIS. This chapter is designed to provide some basic information on doing just that.

Chapter 9

Sharing Geographic Information across Organizational Boundaries

As development of GIS progresses and technical problems are overcome, new problems arise. One of the frequently mentioned problems is that of the need for information and databases that may be housed in several organizations. In this chapter we explore the organizational and managerial roots of difficulties in sharing databases. Historically, a combination of an organization's structure and operations have resulted in the fragmentation of work projects into individual tasks and growing powerlessness among manual and clerical workers. In contrast, such fragmentation of tasks tends to increase the power of knowledge workers and their organizations since such workers and organizations often control information that is valuable in that it is unique and indispensable. In this chapter we identify three means by which alliances may be formed and information may be shared: appeals to professionalism, coercion, and bargaining. We propose a theory based on the relative power of the participants to predict which of these three strategies will be used. Finally, we propose a three-stage conceptual model that delineates (1) a set of facilitators of information sharing in an organization, (2) the process of information sharing, and (3) consequences of enhanced information sharing.

With the steady rise in acceptance of GISs in recent years, there is general agreement that the success of organizations in both the public

---

Portions of this chapter were from "Information Sharing in an Interorganizational GIS Environment" by Z. Nedovic-Budic and J. K. Pinto (2000), *Environment and Planning B: Planning and Design*, 27, 455–474. Adapted with permission from Pion Limited, London.

and the private sectors can be greatly enhanced by the open exchange of geographic information across organization boundaries. Certainly, from a technical perspective, data are now far easier to share owing to their electronic form. However, the rapid increase in the number of organizations adopting GIS technology has belied the fact that between and among organizations there has been a general inability, and often unwillingness, to share data and information across organization boundaries. The waste caused by duplication of effort—which is due largely to lack of information exchange among local, state, and federal governments and the private sector—remains a significant impediment to the more effective and efficient use of GIS throughout society.

Many different strategies for managing information sharing have been proposed and to some degree adopted. For example, some public- and private-sector alliances have attempted to develop data warehousing initiatives, in which a central repository for all geographic data is established and overseen by some governing body. Typically, warehouses are set up in which partner organizations each contribute their own data and expect to be able to draw upon other available data that has been contributed by partners. Interestingly, though a seemingly equitable means for ensuring that all cooperating agencies have equal access to the information, studies and anecdotal evidence suggest that the success rate of information sharing through warehousing is mixed (Nedovic-Budic, Pinto, & Warnecke, 2004). While some partnerships have enjoyed good success with their warehousing initiatives, in other relationships a variety of coordination issues such as cost sharing and location of the data have led to some agencies deliberately withholding data, "backdoor" politicking, and conflict.

It is ironic that at a time when the means for data sharing are at an all-time high, the motivations are still often lacking. The widespread use of the Internet and better established standards for joint ownership of data would suggest that sharing geographic information should be a relatively simple process in which interested parties can communicate with each other and bargain for or simply purchase this data. Interestingly, it is becoming increasingly clear that the chief problems with information sharing are behavioral rather than technical; that is, while technological advances make it relatively easy to share data, organizational and personal impediments continue to make data sharing a significant challenge (Higgs, 1999).

In this chapter we seek to examine some of the roots of the problem of gaining cooperation across organization boundaries in sharing geographic information. We suggest a hypothesis to predict under what circumstances organizations will employ various strategies to effect information sharing, based on their power relative to that of other organizations involved. We also propose a conceptual framework that addresses the influence of a set of antecedent constructs (accessibility, incentives, superordinate goals, bureaucratized and formalized rules and procedures, the quality of exchange relationships among organizations, and resource munificence/scarcity) on the attainment of both cooperation among organizations and optimal use of GIS information. Finally, we report on the results of some important research investigating the motivations behind interorganizational data sharing (i.e., some of the most common reasons why organizations find it useful and appropriate to share geographic information across organizational boundaries). The results of this research offer important implications for GIS professionals who are interested in developing similar "open-access" agreements but know little of the most salient mechanisms by which to initiate sharing relationships.

Information-Sharing Alliances

There are several bases for the formation of alliances, or "intergovernmental systems," as they are sometimes called. Olson and Zeckhauser (1966) and McGuire (1974) suggested that even when people (or organizations) are devoid of feelings (either positive or negative) toward one another, they may find that it is in their interest to organize for the purpose of providing collective goods. Among the bases for the formation of alliances are professionalism, coercion, and bargaining.

Appeals to professionalism may sometimes be a motivating factor in the development of alliances that can facilitate interagency information sharing. Milward (1982) and Keller (1984) noted the importance of functional interests and professionalism in establishing intergovernmental systems. Gage (1984: 136) argued for the importance of understanding such networks as "instruments for establishing and maintaining political networks to accomplish policy objectives." In some cases, professionals may respond to a sense of professionalism, putting aside interagency rivalries to pursue a common goal. Such short-term sublimation of individual and agency goals to address a larger picture can sometimes result in long-term benefits aside from the achievement of a specific common goal. Such joint ventures may stimulate employment opportunities in the field and foster consistent support for the agencies in specific policy areas.

In some instances, the development and maintenance of alliances will be far more difficult, depending on the negotiation of an acceptable exchange among the parties involved. As Weber (1968b: 73) asserted, "Rational exchange is only possible when both parties expect to profit from it or when one is under compulsion because of his own need or the other's economic power."

In general, intergovernmental systems (or "networks," as they are sometimes called) are characterized by an uneven distribution of power (Keller, 1984; Lindahl, 1919/1958; Milward, 1982). According to Milward (1982), within this environment of uneven power distribution, it is not unusual for factions to compete for power to assure that their goals are ultimately adopted and implemented by the intergovernmental system as a whole (470). The result is an ongoing search for equilibrium among the members of the system (see Keller, 1984; Milward, 1982). Equilibrium is not a static condition, but a process outcome that the members of the system achieve through their efforts to gain power. Inasmuch as the quest for power is ongoing, conflict arises as the equilibrium point for the system as a whole changes.

While it seems to be inevitable, conflict should not automatically be viewed as a negative element (Buntz & Radin, 1983; Pondy, 1967). Rather, conflict is the means by which the intergovernmental network achieves an equilibrium of power. In some instances, conflict can produce negative effects; in others, it serves an important integrative function within the network (North, Koch, & Zinnes, 1960; see also Pondy, 1967). Keller (1984) noted that as a means of improving their power position within the network, members of the network may attempt to link their own organization missions with the values held by powerful external groups. Pondy (1967: 313) shares this view: "A major element in the strategy in strategic bargaining is that of attitudinal structuring, whereby each party attempts to secure the moral backing of relevant third parties."

A Theory of Information-Sharing Strategies

The literature on alliances suggests three separate ways in which interagency alliances occur: (1) appeals to professionalism, (2) coercion, and (3) bargaining. Appeals to professionalism may in some cases represent an appeal to somewhat altruistic noble values. In other instances, such appeals may reflect crass self-interest on the professional level. An appeal to professionalism has as one of its advantages its very low cost (i.e., "Talk is cheap"). It is therefore readily available to any organization.

The second means by which interagency alliances occur is through coercion. In some instances, coercion comes by way of controls placed on one level of government by some more powerful level of government. For example, physical development projects of a certain size that are proposed as federally funded efforts are subject to the terms and conditions of the National Environmental Policy Act of 1969. Similarly, state governments may have the authority to require specific information from local governments.

In interagency networks where the power structure is less well defined, or where there is minimal difference in the power of the various agencies, coercion may be impossible. In these instances, bargaining appears to be the most likely means of achieving an agreement on the information (Azad, 1997).

Within the basic concept of bargaining, organizations have a variety of resources at their disposal. In some instances, information swaps may be possible. Some organizations may have the economic resources to purchase information from other agencies or to provide some other tangible economic considerations.

Two factors stand out as central to achieving agreements on information sharing: (1) the value of the information to the negotiating agencies and (2) the interagency power structure. Assessing the value of information continues to be a nagging problem. Information does not have value in and of itself, but rather its value is related to its utility to its potential users. One clear indication of the value of information is the price it commands in the market. The very existence of information brokerage firms provides evidence that market methods of valuation do occur. However, assessing the value of information held in public or private sectors remains an inexact art, at best.

We argue that it is possible to identify which of the three types of resolution will occur—appeal to professionalism, bargaining, or coercion—on the basis of the balance of an organization's power. Figure 9.1 identifies under which power structures each of the types of resolution will obtain.

This model assumes that agencies will seek the least-cost resolution. Within this model, the two least-cost resolutions are coercion and appeals to professionalism. Coercion, however, is available only to organizations that possess the power or authority to pursue it. Appeals to professionalism are available to everyone. Where the balance of power favors the seeker of information, that organization may exert its authority and

Owner of Information

		Powerful	Powerless
Seeker of Information	Powerful	Bargaining	Coercion
	Powerless	Appeal to professionalism	Bargaining

FIGURE 9.1. Distribution of power and strategies to achieve information-sharing alliances.

demand the information from the weaker owner of the information. Where the balance of power favors the owner of the information, the weaker seeker of the information has neither the authority at its disposal to demand the information nor the power needed to enter into a bargaining situation. When the seeker of information is relatively weak, it must rely on appeals to altruistic notions of professionalism and the public good. Bargaining can occur only when both the owner and the seeker of information possess roughly equivalent power, although we would suggest that it makes little difference if both are relatively powerful or relatively weak.

The presence of complicating factors, such as the value of the information and the relative power of the agencies involved, gives rise to uncertainty about the resolution of specific cases. Indeed, the relative value of the information in question to the agencies involved is likely to become part of the power equation. Again, because of the difficulty of accurately assessing the value of information, this contribution of the value of information to the relative power of the agencies is unknown. Empirical research is needed to ascertain the validity of the model developed as Figure 9.1 in this chapter. Actual case studies of information sharing will be needed to gather this information. Case studies should yield valuable information about the nature of the interagency bargains agencies adopt to make possible the sharing of information needed for larger scale GIS.

Antecedents and Consequences of Information Sharing

In addition to examining some of the means by which interagency alliances are developed, it is our goal to propose a research framework for the study of facilitators of information sharing among organizations. Within the organization theory literature, the area of interorganization cooperation is one that has seen far less research to date than intra-organization cooperation. Furthermore, little research has examined the types of factors that can have a positive effect on convincing organizations to share information. Our goal is to offer a framework for understanding information sharing based on the expected effects of a variety of factors in organizations regarding cooperation.

In attempting to address the issue of GIS information sharing among organizations, it is first necessary to provide a context within which such information is often needed. We propose three distinct contexts requiring interorganization GIS information sharing. The first information-sharing context is situation-specific or project-driven. In this situation, two or more agencies come together to work on a common problem that both parties need to address. The two organizations bring their own data and expertise to the table and share information with each other in an effort to successfully solve the problem. For example, hazardous waste disposal is an environmental problem that may require the cooperation of several federal agencies with overlapping responsibilities. In order to develop a comprehensive and effective solution to a hazardous waste problem, several agencies come together to share GIS information that they each possess in order that all parties can contribute to an optimal solution. As this example illustrates, the problems associated within this context are usually nonroutine and nonrecurring, requiring that they each be addressed on a case-by-case basis with their own distinct solutions.

The second information-sharing context is one in which different agencies may be addressing completely different problems but have a need for very similar information. Because they have needs for the same information, organizations develop procedures by which they can regularly share and exchange information with each other. For example, perhaps the Bureau of Indian Affairs and the Bureau of Land Management determine that they have overlapping needs for information about the same federal lands. This need for common information encourages collaborative (sharing) behavior. Before they begin to exchange information, an analysis is performed to determine the procedures (both technical and administrative) by which these two bureaus can most effectively exchange GIS information. In this context, often the first step in the sharing process is to determine some sort of protocol regarding mutual access to either party's information. The focus is on the needs of the cooperating parties. Meeting those needs might result in either a case-by-case approach or a long-term data-sharing protocol. However, the protocol established to meet the needs of the participating parties may be highly inappropriate for effective sharing with additional parties or the broader community.

The third context for information sharing among organizations is based on developing standardized or generalizable patterns of exchange. In this context, information is readily accessible to all parties and can be accessed in useful forms from a central location, such as a data bank, or from networked or decentralized sources. Organizations simply engage in a routine sharing process through the central storage facility or network of decentralized facilities when they need information.

While all three contexts for information sharing have validity, for purposes of illustrating our framework, we have chosen to focus on the first context for information sharing (i.e., that of organizations working together in an ad hoc manner to solve a specific problem). Our primary reason for selecting this case lies in its representation of a common aspect of the geographic information sharing problem. Readily evident and observable are the many situations that require two or more agencies to exchange GIS information in order to solve a problem that the agencies have a common interest in. As a result, in focusing on this more popular context, we are addressing concerns that are immediate and compelling to a wide range of individuals charged with the task of attempting to develop better methods for information sharing. Furthermore, as a result of gaining additional knowledge into the facilitators of organization information sharing, practitioners and researchers will likely be in a better position to gain insights for those investigating issues of information sharing within the other longer term contexts articulated above. So many technological and infrastructure issues must be addressed in tandem with the interorganizational issues that any research program will necessarily need to be active and malleable over time.

This section of the chapter focuses on a three-stage conceptual model that identifies both antecedents and consequences of interorganizational cooperation in sharing GIS information and technologies. Research in the areas of organization power and political behavior (Pfeffer & Salancik, 1978), channel relationships (Dwyer & Oh, 1987), negotiation (Wall, 1985), and intergroup conflict (Thomas, 1976; Walton & Dutton, 1969) have sought, as one of their goals, an increased understanding of the factors by which improved cooperation can occur. While this research has led to an enhanced understanding of cooperation from an organization theory perspective, an analysis of the factors that can lead to greater sharing has not been attempted in organization environments with a focus on the dynamics of sharing geographic information. Furthermore, the relation between GIS information sharing, on the one hand, and organization effectiveness and efficiency, system satisfaction, and decision making, on the other hand, has not received attention to date.

Previous research on cooperation and information sharing has focused almost exclusively on intraorganizational collaboration—that is, attempting to better understand how different functional groups within a single organization can develop more cooperative relationships. This research stream—best represented by the work of Lawrence and Lorsch (1967); Gupta, Raj, and Wilemon (1986); and Souder (1981, 1988)—empirically examines relationships within an organization, usually between specific functional groups, and suggests factors that are important in fostering these relationships. The findings deal with the effect of interdependence, or resource dependency, and coordinating mechanisms (i.e., formalized rules and procedures) on cross-functional interaction. In addition, the research results suggest that the similarity of functional departments—as far as duties and objectives are concerned—positively influences the amount and type of communication between the departments. Finally, this research has been very useful in offering prescriptive advice on methods to better facilitate information sharing among different functional departments.

What have been missing from the literature are systematic attempts to develop a framework of antecedent variables that can improve the likelihood of creating positive collaborative relationships between organizations. Specifically, little is known, for instance, about the reasons for governmental agencies and other GIS-using organizations sharing or not sharing GIS-related information. There is strong evidence to suggest that considerable duplication of effort occurs across organizations because of a basic lack of will to cooperate. Although the technical ability to share geographic information might be readily accessible or achievable, the incentives for an organization or a person to share are insufficient to overcome the impediments. In other instances, data and information are held closely as sources of control and power, even within government offices. In these instances, there is often outright unwillingness to share this "proprietary" information.

Little is understood about the factors that can act as facilitators, or antecedents, of information exchange among organizations. This section draws on literature from a number of sources, including organization theory, intergroup dynamics, exchange theory, and political economy research, in order to posit a model of interorganizational information sharing.

A Conceptual Framework

A variety of factors can act as facilitators or inhibitors of information sharing and cooperation among organizations. These factors range from such individual variables as the personalities of group members, use of information as a source of personal power, "turf" battles, interpersonal relations, and training and skills (Johnson, 1975; Kelly & Stahelski, 1970; Pavett & Lau, 1983; Pfeffer & Salancik, 1978; Schmidt & Tannenbaum, 1960) to such organization variables as political concerns, reward systems, and cultural norms (Lawrence & Lorsch, 1967; Moch & Seashore, 1981; Mintzberg, 1979; Pfeffer, 1982; Shapiro, 1977). In addition to these environmental constructs, the state of computing management for each organization (Kraemer, King, Dunkle, & Lane, 1989) could be investigated and treated as antecedents of information sharing. We propose a conceptual framework to address the expected relationships between a variety of these antecedent variables and the attainment of interorganization cooperation, as well as investigating the "second-level" effect of interorganization cooperation on projected consequences, or outcomes, of information sharing. These consequences are usually assessed as increased efficiency of' organization tasks, increased effectiveness of output (Gillespie, 1991), and information system and partner–organization satisfaction (Ives & Olsen, 1983).

The antecedents shown in Figure 9.2 were chosen for several reasons. First, we felt it was necessary to identify two types of antecedents of information sharing and interorganization cooperation: those that are to some degree under the control of the project team members and those that reflect larger, organization-level constructs. Ease of communication, accessibility, and rules and procedures are proxies for organization variables that significantly influence associations among individuals and encompass many implicit and explicit aspects of an organization's strategy, structure, and culture (Peters, 1990; Pinto, Pinto, & Prescott, 1993). For example, organizations that (1) permit their GIS departments to associate with other parties at their locations, (2) modify their work schedules to meet the demands of the project, and (3) develop their own rules and procedures to facilitate cooperation represent a different type of culture

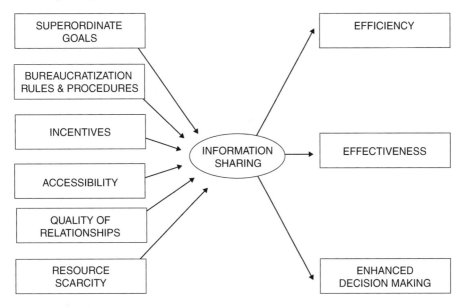

FIGURE 9.2. Antecedents and consequences of information sharing.

than those organizations that do not permit the same degree of latitude. Furthermore, the acceptance of superordinate goals by team members should transcend and mitigate the role of individual factors. Additionally, there has been little organization-related field research utilizing superordinate goals as a method of facilitating cooperation. The quality of exchange relationships argues that high levels of trust and minimal opportunism mark positive interorganization relationships. Finally, research on resource munificence argues that organizations that operate under conditions of increasing resource scarcity are more likely to regard internally generated information as a source of power to be held over other parties. As a result, according to Pfeffer and Salancik (1978), there is a strong tendency to hoard, or refuse to share, that information. The establishment of means for sharing information among organizations is a nonroutine task, whereas the goal is that actual sharing should become routine. Our choice of antecedents represents the need to balance past research findings that have dealt primarily with teams within an organization unit, on the one hand, with our objective of studying cross-functional teams working on nonroutine projects, on the other.

Figure 9.2 illustrates the conceptual framework that identifies a three-stage path analytic model delineating the factors argued to facilitate interorganization GIS information sharing. In the following sections, we examine the importance of each of these factors, the importance of cooperation itself, and perceived outcomes, or results, of enhanced GIS information sharing.

The Need for Information Sharing

The responsibility for many project-based activities often overlaps, or is held jointly, between two or more GIS-using organizations. It has long been acknowledged, from an organization theory perspective, that these "difficult-to-assign activities give rise to such interdepartmental issues as cooperation, coordination, conflict, and struggles for power" (McCann & Galbraith, 1981: 60). To manage the development and implementation of these activities, a variety of integrating mechanisms have evolved, including task forces, liaison roles, and cross-organization teams (Galbraith & Nathanson, 1978; Lawrence & Lorsch, 1967; Mintzberg, 1979). These teams or task forces allow for lateral contact between multiple organizations (Dumaine, 1990). They tend to be temporary groups that exist for the duration of the designated activity. Because of a general unwillingness of some organizations to cooperate and share information willingly, the activities of these temporary task forces have become increasingly important for the long-term viability of an organization. Consequently, efforts must continually be made to develop policies and mechanisms that promote, rather than inhibit, cooperation across organization boundaries.

Antecedents of Interorganizational Cooperation

Superordinate Goals

Classical organization theory originally established the importance of goals in organizations (Simon, 1964). Since that time much has been written on the concept of a goal for an organization (Kono, 1984), the purposes served by goals (Quinn, 1980), the multiplicity of goals in organizations (Raia, 1974), and the hierarchical nature of goals (Galbraith & Nathanson, 1978).

Every organization—and, indeed, every manager—has more than one goal that guides its activities and actions. In theory, different organizations performing similar or complementary functions should possess complementary goals that are derived from a set of general, cross-organization objectives. In practice, however, overall goals are often broken down into specific objectives that conflict with, rather than complement, each other. Consequently, in order for one organization to achieve its goals, another may be required to sacrifice, or at least compromise, its primary goals. Newman (1988) cautioned that a department's goal must be compatible with the activities of related departments so that it will not undermine the results of those activities or make them much more difficult. Consider, for example, the common conflict between state and local or state and federal agencies over GIS data sets. Because the goals of these agencies are different and may often conflict, willingness to go out of one's way to share information collected by someone else is likely to be lacking. This situation results in a strong potential for duplication of data collection and maintenance efforts and may result in the implementation of systems that are each underutilized by their respective organizations. Members of each organization will argue the need for their own system and are often loath to cooperate or make available readily accessible information that may be needed by other agencies. Aware of such conflict, organizations and oversight groups are continually looking for ways to develop goals that can increase, rather than detract from, information sharing and interorganizational cooperation.

One important, but often overlooked, type of goal in the study of cross-functional cooperation is a superordinate goal. As conceptualized in this chapter, *superordinate goals* refer to "goals that are urgent and compelling for all groups involved, but whose attainment require the resources and efforts of more than one group" (Sherif, 1962: 19). It is important to note, however, that superordinate goals are not a replacement for other goals that the various organizations may have; rather, they are an addition to existing organization goals. Through various controlled field and laboratory experiments, Sherif compiled impressive evidence indicating that when a series of superordinate goals are introduced into a conflict situation, intergroup conflict is reduced and cooperation is achieved. The essence of Sherif's theory is that competitive individual goals cause intergroup conflict but that superordinate goals give rise to intergroup cooperation that enhances group output. Specifically, when groups associate under conditions embodying shared goals or common purposes, they tend to cooperate as they work toward common goals. Because of this cooperation, system satisfaction is enhanced; thus, evidence suggests that a two-stage process is present. Superordinate goals are assumed to lead to enhanced data sharing, which, in turn, is expected to result in an improvement in an organization's efficiency, effectiveness, and decision making.

Organization research on superordinate goals has tended to be conceptual in nature (Stern & Heskett, 1968) or has been conducted in experimental settings in which superordinate goals are manipulated (Johnson & Lewicki, 1969; Stern, Sternthal, & Craig, 1973). Therefore, there is strong evidence to suggest that superordinate goals can have a powerful effect on interorganizational cooperation. What superordinate goals would cause numerous federal, state, and local government agencies and private sector organizations to work toward geographic information sharing? Would such a superordinate goal be achieved by developing the concept of a nationwide library reference system for spatial data and by providing a distributed network infrastructure allowing ready transfer of large data sets? What lesser superordinate goals would enhance abilities as well as motivations to share?

Bureaucratization: Rules and Procedures

"Rules and procedures" refer to the degree to which activities or tasks on a project team are mandated or controlled. According to Galbraith and Nathanson (1978), rules and procedures are central to any discussion of interorganizational cooperation because they offer a mechanism for integrating or coordinating activities, particularly those activities that cut across agency or department boundaries. Early organization theorists such as Taylor (1911), Fayol (1929), and Mooney (1947) relied on rules and procedures to link together the activities of organization members. They posited that coordination could be accomplished by simply establishing rules and procedures throughout the management hierarchy. Subsequent researchers (e.g., Gouldner, 1954; March & Simon, 1958; McCann & Galbraith, 1981) also discussed rules and procedures as a technique for coordinating activities, controlling behaviors, and maintaining the structure of an organization. At the department level, McCann and Galbraith (1981) described rules and procedures as the most common structural variable for assigning duties, evaluating performance, and minimizing the occurrence of conflicts between departments. Providing empirical support for this assertion, Reukert and Walker (1987a, 1987b) found that written or formalized rules and procedures have a significant positive relation to the perceived effectiveness of interdepartmental relations.

Galbraith and Nathanson (1978) noted that rules and procedures can be an effective method for achieving coordination between organizations. Specifically, they argued that standardization through rules and procedures is a useful integrating mechanism only when organizations operate under conditions of relative certainty and routine tasks. As an organization's design becomes increasingly complex, however, the effectiveness of rules as a coordinating device among departments decreases.

It is important to distinguish between the *concept* of bureaucratic control and the *effects* of bureaucracy on information exchange. This alternative assessment of bureaucratization argues that as bureaucracies evolve, they tend to become stifling and territorial and will actually inhibit the flow of information across organization borders (Obermeyer, 1990a). The argument points to an organization's bureaucracy as one of the principal culprits in preventing free information exchange. This position is supported in the management literature by research within a marketing context that has found bureaucratic structuring to be damaging to exchange relationships and to exacerbate opportunism between organizations (John, 1984). While these arguments have merit, they do not negate but rather serve as a complement to our analysis of bureaucratization. We are here examining the concept of bureaucratization from the perspective of control mechanisms (rules and procedures) rather than directly addressing an organization's bureaucracy and its potential for non-cooperativeness. In other words, when one agency sets up a series of procedural steps to ensure that its personnel will cooperate with other agencies, they are using a bureaucratic form of control. This is not to gainsay the potential negative effects that a large bureaucracy can have on cooperation between organizations but rather to argue that bureaucratization may also be viewed from another, more positive perspective.

A second conclusion relevant to our model is that rules and procedures have a means–end interrelation. While rules and procedures are developed ultimately to facilitate the accomplishment of desired ends, they also provide a means to establish cooperation among the individuals or departments charged with a particular task. The degree of rules and procedures is related to the degree of formalization in an organization. Recent studies by Deshpande and Zaltman (1987) and John and Martin (1984) found that increased formalization had a positive effect on the flow of information. Building on this research, Moenart and Souder (1990) suggested that increased formalization between departments produces a more harmonious climate. Therefore, it appears that rules and procedures should have a direct influence on the development of information sharing and cooperation between organizations. Thus, if the role of rules as the means does not facilitate cooperation, then the end state will most probably suffer.

Incentives

Another logical facilitator of an organization's willingness to share information with another has to do with perceived incentives. This argument captures the old question "What's in it for me?" that individuals and organizations frequently ask before engaging in any type of personal or professional commitment. Incentives suggest that an organization or its key members must perceive a payoff arising from the act of cooperating in an information exchange relationship. Such a payoff may be in the form of creating a future bond of obligation or gaining some form of strategic or monetary advantage over rival organizations or agencies.

Under the incentive system, the willingness of one organization to participate in an information exchange lies in direct proportion to the other organization's providing the first organization with some scarce or necessary resource that it does not possess (e.g., money, access to important information). When that other organization communicates its willingness to develop an economic exchange relationship, the first agency must determine if the incentives are valuable enough to warrant the exchange of information for that resource. If the answer is yes, the information exchange will occur.

Accessibility

Accessibility usually determines the type and frequency of associations that occur between organizations. In our model, accessibility is defined as an individual's perception of his or her liberty, or ability, to approach or communicate with another individual from a different organization. Factors that influence the type and amount of association that occurs between organization members include an individual's schedule, position in an organization, and out-of-office commitments. These factors often affect the "accessibility" among organization members. For example, consider a setting in which an individual from a local government is physically located near an individual from a state agency. While these individuals are in close proximity to one another, they may rarely associate because of different work schedules, varied duties and priorities, and commitment to their own agendas. Within a manufacturing organization, Souder (1981) demonstrated that the lack of communication, lack of appreciation, and distrust that often exist between marketing and research and development units is fostered by normal time pressures, work deadlines, and some imbalance of power and prestige. These factors often lead to a perception of "inaccessibility" among the individuals involved. Building on a set of studies, Zaltman and Moorman (1989) found that associations between organization members enhance trust. Associations are easier when parties are accessible. Zaltman and Moorman suggested a causal link between associating and trust. Communication strategies and technologies that can help overcome lack of ability to approach another caused by differences in schedules, position in the organization, and out-of-office commitments should help build trust among people and organizations with the potential for sharing geographic information. In our model, previous research suggests that accessibility should directly affect the degree of cross-functional cooperation.

Quality of Exchange Relationships

Research supports the view that the quality of exchange relationships with potential exchange partners represents a significant criterion for evaluating cooperative information flow (Hunt & Nevin, 1974). The arguments underlying this issue suggest that member expectations of cooperation and information exchange are crucial elements in maintaining quality channel relationships. If these elements are characteristics of the system, different GIS-using organizations would be expected to fulfill obligations and would expect each other to desire coordination; high levels of satisfaction and morale should follow.

Following research by Dwyer and Oh (1987) that was originally compiled to address relationships among member organizations within marketing channels, we analyze the concept of exchange relationships as consisting of three key dimensions: *satisfaction*, *trust*, and minimal *opportunism*. Satisfaction with an exchange partner is a significant issue when evaluating information-exchange relationships. Member satisfaction includes all characteristics of the relationship that the focal organization finds "rewarding, profitable, instrumental . . . or frustrating, problematic, inhibiting" (Reukert & Churchill, 1984: 227). Likewise, trust refers to a party's expectations that another desires coordination, will fulfill obligations, and will pull its weight in the relationship (Anderson & Narus, 1986). When one organization has built up a level of comfort and trust with another party, it is willing to engage in a more open exchange and sharing of relevant information (Deutsch, 1958). This argument suggests that the past history of working relationships will have an effect on future levels of trust and willingness to cooperate. Finally, opportunism is exemplified by distortion of information, failure to fulfill promises, and unwillingness to share what may be considered proprietary information. In effect, as one researcher suggested, opportunism is "self-interest seeking with guile" (Williamson, 1975: 26). When the condition of opportunism is prevalent between two agencies or organizations, there is little likelihood that needed information will be shared between the parties, as each or both of the exchange members view information as a base of power to be held over the other.

Resource Scarcity

The concept of resource scarcity refers to the idea that organizations must contend with a limited pool of resources when conducting their activities. Under certain circumstances, the resource level that is available to an organization may not be constraining but may allow a wide range of activities and options. This state is referred to as munificence. More often, however, organizations are constrained by limited budgets and the availability of technology, trained personnel, and other needed resources.

Research has long found that organizations and agencies operating under conditions of resource scarcity often tend toward the desire to maintain some form of control over other companion agencies. Indeed, the resource-dependence model of an organization's power argues that a method by which one organization can exert control over another is through creating and stockpiling some resource that is scarce and is needed by another organization (Pfeffer, 1982; Pfeffer & Salancik, 1978). In the case of the current discussion, the needed resource may be information of the sort provided by GIS technologies. Because information is viewed as a form of power, the agency that possesses needed information is less likely to share it with another party for fear of losing some base of power in the exchange relationship.

On the other hand, when conditions of relative munificence operate, there is less emphasis placed on hoarding resources, either material or informational. Thus, the environmental condition of munificence offers an increased likelihood of information exchange, while perceived resource scarcity is likely to create the opposite effect, in which parties possessing information are not as inclined to make it readily available to other organizations.

Information Exchange

Many concepts exist that encapsulate the meaning of cooperation between organizations. Among the various terms used to describe the notion of individuals working together to accomplish a specific task are (1) "coordination" (Argote, 1982; Van De Ven, Delbecq, & Koenig, 1976); (2) "collaboration" (Trist, 1977); (3) "cooperation" (Schermerhorn, 1975; Sherif & Sherif, 1969); and (4) "integration" (Gupta, Raj, & Wilemon, 1986; Lawrence & Lorsch, 1967). The variability in terminology raises a critical question regarding what, if anything, the underlying concepts have in common. While each of the terms has a separate and distinct name, each refers to a similar and overlapping idea as evidenced by the commonalities in the definitions. The lowest common denominator that integrates the four concepts is *joint behavior toward some goal of common interest.* For purposes of this research, organization information sharing and cooperation are conceptualized as the degree, extent, and nature of interpersonal relationships among member from multiple organizations.

The need for information sharing and cooperation stems from the complex interdependencies among members of different organizations charged with complementary objectives. As Thompson (1967) suggested, greater interdependence requires a greater cooperation effort. Unfortunately, problems associated with cooperation between organizations result not only from the interdependence of work process and technology but also from conflicts over authority and jurisdiction among team members representing different departmental units. Thus, information sharing within an organization is essential in the implementation of decisions. It has been shown to promote productivity by helping individuals perform more effectively (Laughlin, 1978). Laughlin argued that organizations that are cooperating tend to (1) understand and be influenced by each others' interests and ideas, (2) seek and give information, (3) communicate about tasks, (4) more readily assist each other, and (5) rely on division of labor.

The previous discussion has established a link between antecedents and cooperation between organizations. Research has also shown that the antecedents may not have a strong direct link to project performance. Furthermore, GIS information exchange appears to be directly related to the outcome measures of increased efficiency and client satisfaction. Thus, our model of cooperation between organizations mediates between the antecedents and outcomes measures.

Consequences of Cross-Functional Cooperation

As Figure 9.2 indicates, greater sharing by organizations of geographic information is not simply an end to be sought for its sake. Rather, we have posited a three-stage flow process in which information sharing represents the mediating link between the antecedent variables and outcomes in organizations. In effect, our model suggests that enhanced information sharing will lead to several desired outcomes that various organizations seek: efficiency, effectiveness, and improved decision-making ability. In addressing each of these outcomes, we will be making use of previous research in the area of the measurement of GIS use and effect developed by Gillespie (1991) and Zwart (1991).

One expected outcome of enhanced information sharing between organizations is greater internal efficiency of operations for each agency. When one organization is able to make direct contact with another party that possesses needed information, there is far less likelihood of replication of effort in creating databases. An organization's efficiency is enhanced through this sharing process. Increased efficiency from information sharing is measured, according to Gillespie (1991), by taking the difference in variable costs between the alternative methods for producing the desired output. In the case of geographic information, an estimate of the resources (labor, time, and money) needed to develop required GIS information in-house would be compared to the cost of retrieving such information from external sources with whom the organization has developed a cooperative relationship. If the difference is positive, then it could be argued that information sharing has resulted in increased efficiency.

A second outcome of information sharing is that of increased effectiveness in an organization. Effectiveness has been defined by Gillespie (1991) as the case where GIS "increase the quality of the output or produces a new output" (A-85). Gillespie has further suggested a three-step process for measuring effectiveness benefits. First, it is necessary to determine how the GIS output is different. In other words, what is being used or produced now that was not operational or available before? Second is the need to determine what effect each of the changes in GIS output has on the users of that output. One way to assess the changes that can take place is to examine the activities of organization members. Given the availability of new sources of information through cooperation between organizations, do members make use of this available information? If the answer is no, then it is likely that sharing has not increased effectiveness. Finally, the third step in assessing effectiveness is to determine the value of each of the effects on the users. Obviously, this step is the most complex and difficult because it requires that some figure of value be assignable to the set of effects.

The final source of outcome to assess is the effect on decision making of the new cooperative arrangement between organizations. Zwart (1991) argued that unless we can determine that utilization of geographic information has led to enhanced or better decision-making capabilities, its effect is minimized. Lucas and Nielsen (1980) reinforced the importance of improved decision making by suggesting that we need to assess effect by examining the degree to which information is utilized. They argued that for information to be fully utilized by an organization, it must not simply be referred to in decision making, but it must actually lead to changes in an organization's values or in managerial decisions. As a result, if information that is shared between organizations does not lead one party to reassess decision priorities or value structures actively, the third criterion of outcome (decision making) has not been fully addressed.

Motivations for Information Sharing: Research Findings

Zorica Nedovic-Budic and Jeffrey Pinto launched a study to empirically investigate the primary motivations for and mechanisms by which organizations share geographic data with each other (Nedovic-Budic & Pinto, 2001; Nedovic-Budic, Pinto, & Warnecke, 2004). Starting with a set of cases of city and county governmental agencies engaged in data-sharing initiatives, they conducted a series of in-depth interviews and analysis to determine the reasons behind these independent agencies' decision to share information and the benefits they have perceived to be derived from data sharing. In this section, we consider some of their significant findings, drawing the obvious links to general theory of interorganizational information sharing.

The first significant issue to be addressed asked the fundamental question: Why share information?

Significant Issues

Table 9.1 summarizes the issues that were identified as significantly affecting the interorganizational efforts to jointly develop a GIS and to share or exchange geographic information across the five case studies. The issues are grouped into two categories: those unique to the coordination process, and those relevant to the GIS implementation process. Coordination issues were fundamental for the GIS and database sharing activities. The research indicates that the nature of the coordination process was, in fact, the key to establishing an atmosphere of trust and mutual collaboration and for the overall success of each multiparticipant project. Even in the

TABLE 9.1. Coordination and Implementation Determinants of Geographic Information Relationships

	Northwest	Southwest	Midwest	Southeast	Northeast
Coordination factors					
1. Contributions	(+)	(−)	(−)	(−)	(+)
2. Control	(+)	(−)	(−)	(−)	(+)
3. Negotiation and persistence	(+)	(+)	(−)	(+)	(+)
4. Commitment	(+)	(+)	(−)	(+)	(+)
5. Authority and stability of leadership	(−)	(+)	(−)	(+)	(+)
6. Database responsibilities	(−)	(−)	(−)	(+)	(−)
7. Data ownership/location	(+)	(−)	(−)	(−)	(+)
8. Access to the data	(+)	(−)	(−)	(−)	(+)
9. Technological change	(+)	(+)	(−)	(+)	(+)
10. Organizational change	(−)	(−)	(−)	(−)	(−)
Implementation factors					
1. Top support	(+)	(+)	(−)	(+)	(+)
2. Long-term funding	(+)	(−)	(−)	(+)	(+)
3. Project scope	(+)	(−)	(−)	(−)	(+)
4. Timing	(+)	(−)	(−)	(+)	(−)
5. Expectations management	(+)	(−)	(−)	(−)	(+)
6. Communications	(+)	(+)	(−)	(+)	(+)
7. Demonstrable progress	(+)	(+)	(−)	(−)	(+)
8. Personalities and private agendas	(+)	(+)	(−)	(−)	(+)
9. Project champion	(+)	(+)	(−)	(+)	(+)

Note. (+) = the issue was addressed adequately or not problematic; (−) = the issue was problematic and not addressed adequately.

system and data-sharing initiatives that have been relatively effective, the difficulties were primarily caused by the unwillingness or inability of the parties involved to adequately satisfy members' concerns regarding responsibilities, equity, and fairness. The participants' attitudes were decisive in determining the level of success in joint GIS and database activities. Among the five cases studied, the northwestern and northeastern projects were advanced more smoothly, to a great degree due to positive coordination attitude and management. In those two cases, however, there was a single major project leader: the county government. The other three cases had a shared leadership between county and city jurisdictions, and, although only in the midwestern case the achievement of joint objectives was considerably delayed, addressing the coordination issues was much more difficult in all three.

While the coordination issues were pertinent and unique in situations where multiple organizations and agencies were involved, the general implementation issues applicable across and within each organization and agency also exerted substantial influence on the outcomes of interorganizational GIS activities. Along with the common protocols being devised and agreed to, a number of important implementation issues had to be addressed. It was clear that overall success depended on each organization's implementation capacity and management (Brown & Brudney, 1993). Consistent with the findings of other GIS implementation studies, the pertinent issues included top management support, secured continuous project funding, well-defined project scope, management of expectations, timing of specific activities and phases, demonstrable progress, and avoidance of personal and political agendas and conflicts. The pattern of impact of particular implementation issues on joint activities was comparable to the one regarding the coordination. The cases with intensive implementation control were able to achieve the objectives of interorganizational GIS and database most effectively; the cases that addressed some implementation issues experienced more crisis situations; and finally, the case with least attention for the implementation process faced major challenges in GIS diffusion and use across organizations involved.

Let us consider in more detail the findings of Nedovic-Budic and Pinto (2000) in terms of the crucial interorganizational coordination issues, including contribution, control, persistence in communication and negotiation, commitment, authority and stability of leadership, data ownership, data access, database responsibility, technological change, and organizational change. Our findings are related to the relevant points in the previous research. The outcomes, costs, and benefits of the joint GIS and database activities are considered last.

Contributions

What would each party in the interorganizational relationship be expected to contribute? It was important for all members that each agency's contribution be determined in advance. All groups naturally have concerns that their monetary or other in-kind contributions for developing the GIS and database would be commensurate with their relative sizes, resources, and needs for the data downstream. Rarely was a single agency willing to shoulder what it felt were undue or excessive expenses in developing the database or other joint components. Generally, there was a real desire across the cases for equity in the use of the data or other common resources. The cases confirmed the risks of "overgrazing" the data, fouling or contamination, and data poaching common to pooled IS/IT resources (Kumar & van Dissel, 1996). Consequently, a great deal of negotiation among the various agencies had to address this issue directly. What would be each partner's expected commitment to the pooled database development and other joint activities and products? What would be each parties' concomitant degree of returns in the form of data, services, equipment, staff, or other benefits?

Control

How would the consortium ensure equal control of the GIS, database-related, and other joint issues? Every interviewed member of the interorganizational GIS initiatives expressed the desire for a fair decision-making process that would ensure that participants have an adequate control of the common activities and an equal partnership in the initiative. Depending on their resources, power, and role in the partnership, however, the organizations differed in their views about what fairness and equity represents. Voting rights and decision authority have to be carefully determined, often through protracted negotiations, prior to having all partners sign off on the protocol. Among the key questions that these negotiations were forced to answer were creating solutions that were perceived as equitable. Not surprisingly, large agencies tended to favor a protocol giving them a level of power proportional to their size and contribution, while the smaller ones were anxious to maintain equal voting rights for all members, regardless of size. Vaguely defined or unsatisfactory control structures were the source of major rifts and problems in two of the cases studied.

Persistence in Communication and Negotiation

One key finding that emerged from the research was the vital need for organizational persistence to make the interorganizational arrangements succeed. Indeed, it was easy and highly tempting for various organizations studied to break away or dissolve the partnership once the difficulties arose or their own needs were secured. For parties not genuinely interested in the joint venture or those that found that the commitment required would outweigh the organizational benefits, it was a matter of finding a good reason and opportunity to discontinue the interaction. The breakaway actually happened in two of the five cases—one before any joint activity was started, and the other after the database was developed and various joint services and applications attempted.

Persistence was the most clearly manifested success factor in the process of endless negotiations over all issues of joint concern. Consistent with Evans's (1995) finding, one of the key challenges in the negotiation process pursued by the organizations studied was the importance of maintaining a coordinated overall focus despite differing agendas and styles among the participants. Pursuance of other political and power agendas while resolving the multiparticipant GIS issues was highly detrimental in two cases. In the remaining three cases, the extraneous issues were early recognized as a threat and either avoided or dealt with directly. Harvey (1997) suggests that identifying semantic differences and commonalties between concepts held by participants and creating a common working language are the initial steps toward successful communication and negotiation of positions, solutions, and ideas. The necessary interorganizational communication evolved at both a formal and an informal level. The research evidence supports the contention that often the "real work" of establishing coordination protocols took place in informal settings rather than in the structured settings.

Finally, coalition building, bargaining, and willingness to compromise as important dynamics in effective interorganizational arrangements were all exercised in the cases studied. These activities have long been recognized as standard activities when multiple parties are involved in a project (March & Simon, 1958).

Differential Commitment Levels

In the cases studied, different groups engaged in interorganizational GIS and database activities with varying levels of commitment. Some entered the relationship with less than full commitment to that process, some partners emerged as project champions, others experienced frustration based on their perception of inequitable resource expenditures and returns, and still others engaged in "guerilla warfare" designed to push private agendas at the expense of the overall project goals. It was true that the agencies involved in interdepartmental relationships (1) lost some of their freedom and (2) had to invest energy and resources to develop and maintain relationships with other organizations. This has been long acknowledged as necessary in interorganizational dynamics (Van de Ven, 1976). In all cases, however, the success of the joint investments of various resources, including financing, time, energy, and staffing, depended heavily on the commitment to the common cause of sharing. True commitment helped overcome many of the obstacles in the process of joint system and database planning and implementation, and also maintained the focus on the matters that were pertinent to the joint activities. The participants who were committed "for the wrong reasons" were ultimately disruptive and sabotaged the attempts to coordinate and find common solutions.

Obermeyer (1995b) notes that one way to enhance the commitment of each participating member to the sharing alliance is through an interorganizational agreement that requires the contribution of both money and actual ongoing effort from each partner. The benefit of this strategy is that it also seems to work to advantage by increasing the stakes of each of the participants from the start. Indeed, in all cases studied, intergovernmental agreements and memoranda of understanding were used to formalize and affirm the participants' commitment. In one of the cases, a formal agreement developed in the early stages of the joint system and database development process was respected even when the parties abandoned further coordination. The agreement kept the channels and mechanisms for database exchange open.

In essence, the key to coordinated GIS and database activities is convincing each member organization of the important synergies that derive from a long-term commitment to the sharing arrangement, not simply for the data's sake, but also to enhance future collaboration at all levels between the organizations. The success, therefore, depends on the spirit of cooperation and commitment to sharing on the part of all members

(Meredith, 1995; NGDPF, 1993). In order for that spirit to emerge, there must be a sense of teamwork, shared understanding, trust, and mutual credibility (Citera et al., 1995).

Authority and Stability in Project Leadership

Authority to act and the stability of project leadership structures were also important findings that emerged from the case studies. Each participating agency needed to perceive that they possessed the decentralized authority to implement their plans. As one interviewee noted, "There is nothing more frustrating than spending a huge amount of time ironing out all the details of the [sharing arrangement] only to have it shelved or shot down by high-level bureaucrats. Either we have the authority to act or we don't." As with any complex venture, participants needed to feel empowered to plan, make decisions, and bring them to realization. When a successful partnership was developed, it was often due to the partners' ability to immediately move ahead with their plan in order to see its impact, resolve any problems, and transition from theoretical planning to realizing practical benefits as quickly as possible. When the main level of interorganizational activity occurred at staff or middle-management level, without direct support or involvement of higher level administrators, the ideas and plans developed in numerous meetings were hard if not impossible to transfer back to the participants' local settings, and support them with resources. Organizational power (i.e., the ability to exert influence and bring about desired outcomes) was directly related to progress in joint GIS activities in all cases studied. The area where the authority was the most critical was in enforcing adherence to local standards and database management commitments. This finding is supported by previous empirical research (Brown & Brudney, 1993; Campbell & Masser, 1991; McCann, 1983).

Database Responsibilities

Who will be responsible for developing and maintaining the geographic information? Along with the question of ownership and actual possession of the geographic information is the issue of the degree to which each partner will be responsible for sharing the costs and duties of data acquisition, data entry, and maintenance. Research findings indicate that interorganizational relationships that clearly spelled out partner responsibilities for the joint database were more likely to be harmonious and experience less rancor or political agenda setting. Several additional issues were found to be very important to the data development and maintenance process. First, it was necessary to identify and attract the most important data providers and enlist their support early in the GIS database relationship. In the local government settings where the parcel is the basic spatial unit of mapping, tax assessor offices were those key data providers. The tax assessors were successfully involved in joint GIS activities only in two of the five cases analyzed. In two cases, alternative sources of parcel data or other feature maps were used as base maps. The midwestern case was stalled until the tax assessor's office gained enough interest to independently pursue digital database acquisition and to provide access to its data, but without much consultation with other agencies. The potential participants often evaluated the value of the partnership based on the involvement of these key actors.

The second issue had to do with securing additional resources to those units that were charged with maintaining the data. In all five cases those units began to incur additional workloads, expenses, and responsibilities as a consequence of their involvement in the partnership. When these units perceived inequities in data maintenance commitments, they were prone to downgrade their own support of the system. In the absence of staffing, funding, equipment, and training provisions, the agencies assigned database maintenance responsibilities were likely to fall behind in timing and quality of database update. They also tended to depart from prescribed standards and procedures as another consequence of the inadequate support for database maintenance duties.

Addressing the assignment and support for database responsibilities is a crucial component of interorganizational GIS activities. The fact that the cost of updating data will ultimately dominate other GIS costs requires the designers of the joint systems to concentrate on the database aspect from the very beginning (Frank, 1992). In the cases studied, the database update was generally well placed in the organizations with compatible functions. Although the database responsibilities were matched with existing organizational missions, the support for the task was rarely secured.

Data Ownership

Who will "possess" the data? An important up-front decision point has to do with determining at what site the data will physically exist. For several

participants across the studied examples, there was a perception that ownership constituted dominant control. Consequently, in these cases a considerable amount of time may be spent ensuring that equal access to the
data will be available regardless of its location. Openness with regard to
data access, minimal proprietary interest in data, and no gains expected
from data distribution are all conducive to less conflict and tension
regarding the ownership of data.

Another key aspect of the possession issue revolves around the location of the coordination or service unit. All parties require that this unit
be perceived as neutral (i.e., having no vested interest in one agency or
organization having greater access to data than others). The participants
openly or privately may question how the location of the coordinator
would affect data ownership or relative position of each of the data-
sharing partners. In three out of five researched cases, the difficulties
experienced by the partnership members stemmed primarily from the
coordinating unit being perceived as nonneutral. In two cases this perception led to deterioration of the GIS and database-related interaction.
These findings reinforce Sperling's (1995) argument that minimal problems in data ownership are key for successful sharing relationships.

Data Access

How can the parties ensure that all have equal access to the data? A frequently expressed reservation, particularly from agencies who perceived
themselves as "junior" partners in the data-sharing initiative, is how to
ensure a sense of equity and fairness in data exchange and access. It is
important for these agencies that formal safeguards be in place to ensure
that jointly held geographic data is also jointly available. One of the most
important lessons derived from established partnerships is the need to
clearly indicate the nature of the sharing structure early in the process.
Simply allowing the GIS and database interactions to evolve over time
without set rules and procedures often only ensures increasingly suspicious partners. The key, as has been noted in previous research, is establishing a stable and simple relationship structure (Brown et al., 1996).

A related issue is often raised concerning the access to proprietary
data, giving one participant a competitive advantage over other partners
in pursuing their organizational mission. Under these circumstances, as
has been proposed by Calkins and Weatherbe (1995), it is important to
maintain the proprietary data outside the core database. Clear and open
up-front specification of any possible restrictions in distribution and use

of the organizational spatial data are necessary for avoiding any potential future misunderstandings and conflicts regarding the proprietary data.

Technological Change

Effects of advancements in networking technology, distributed GIS, and the Internet are obvious. With the rapid increase in distributed computing environments, in all cases studied there was a departure away from centralized systems of data storage. The new configurations had an impact on spatial data processing and sharing relationships because the data was often no longer linked to a central depository, ensuring one party's effective control of data dissemination. Associated with this shift to decentralized data structures and networked environments, in all cases there were tensions between the traditional data-processing departments and staff and the participants in distributed GIS-related activities. In essence, this was the clash of two computing cultures. Early to mid-1990s was the most dynamic time in terms of the changing computing environments and their corresponding organizational implications. In two cases, which initiated their joint GIS and database activities in the 1980s, the change involved a major move from centralized mainframe equipment to workstations and PC platforms and new networking setups. In three other cases (one of which did not materialize until later in the 1990s), the switch was not that dramatic, since the distributed technologies were already available. The process of technological change, however, is ongoing. In one case there was a deliberate timing of the system installation process in order to capitalize on the most recent technological developments. Internet-based access to GIS database or clearinghouses were available or planned for the near future in all but one case.

Several researchers have commented on the impact of distributed GIS. Orthner, Scherrer, and Dahlen (1994) confirmed that the environment of information systems was changing from centralized terminal/mainframe configurations to distributed client/server architectures, with the eventual goal of arriving at a totally distributed system that is optimized for responsiveness, availability, and reliability. Frank (1992) also noted this trend, arguing that while maintenance of a central repository and the sharing of the same data among many users may be the best technical solution, it had not been economically feasible in the early 1990s. Taupier (1995) pointed to the inherent trade-off involved in use of distributed GIS, suggesting that the decision to establish a central clearinghouse for data versus decentralized repositories or a distributed network of data libraries involved balancing the ease of access and the distribution of cost.

The move to distributed GIS carries with it some important organizational implications as well. A number of writers have argued that this technological change requires an equal change in administrative and organizational philosophies. For example, while Azad and Wiggins (1995) have argued that data sharing will be made easier by the developments in IT technology, administrators considering distributed GIS correctly perceive that they will encounter greater "leading edge" complexity due to rapid changes in technology (Onsrud & Rushton, 1995). Finally, as Evans and Ferreira (1995) cogently suggest, these challenges make it vitally important that system implementation plans be flexible enough to accommodate technological changes because of the likelihood that organizational factors will be affected by a changed technological mix.

Organizational Change

Wigand (1988) has noted that new structural configurations can be expected among organizations introducing new information technologies. Indeed, he notes that one result of integrated and distributed data processing is that rigid hierarchical structures are redesigned, resulting in leaner, more flexible and responsive organizations with fewer management levels and more direct information exchange between the top and bottom layers. The case studies revealed no major organizational change prompted by the interorganizational GIS-related activities. In all but one case studied, the coordination was managed through an elaborate system of committees, that ranged in their concerns from policy and strategic perspectives to detailed technical and user issues. The actual organizational restructuring, however, occurred only sporadically. Even when a restructuring was initiated, the interorganizational GIS and database activities were rarely the primary justification for it. The problems of mismatch between new database tasks and procedures and organizational structure were persistent in four of the five organizations studied, suggesting that absent a clear change strategy, a creeping "incrementalism" can simply begin to occur and pose serious consequences on the structure and the efficiency of subsequent operations.

The sense of upcoming change and the uncertainty brought with it was, in fact, unsettling to many agencies and their personnel. One frequent question considered by the participants in the interorganizational relationships studied was what was likely to happen to them and their current operating resource base with the advent of the more free or open access to a larger pool of geographic data. The attempts in managing expectations were predominantly focused on technology. The real concerns, however, were about the implications of the technological change and joint database activities for subsequent organizational realignment.

Previous research shows that the benefits of data integration cannot be fully realized unless the adoption of technology is accompanied by organizational, institutional, and behavioral changes (Alfelor, 1995). In interorganizational GIS activities, it is apparent that those organizational actors who foresee and correctly anticipate the needed changes are more likely to successfully implement their data-sharing arrangements than those who do not foresee such changes.

Benefits versus Costs

The discussions of interorganizational GIS and database sharing ultimately focus on the perceived benefits and costs of such interactions. The traditional view has held that data sharing will enhance organizational efficiency through a reduction in redundant operations, while improving collective decision making as multiple parties are better able to communicate through shared data. Various organizational groups, however, experience different needs (e.g., accuracy, updating frequency) and therefore derive different perceptions of the benefits from data sharing (Sperling, 1995). Nedovic-Budic and Pinto (2000) uncovered some of the tangible benefits that were derived from a willingness to share GIS and databases. Data consistency, enhanced cooperation, and technology transfer to small jurisdictions were among the benefits from joint GIS activities studied in their research. Transaction cost and coordination difficulties were the major costs. These costs and benefits confirm the current state of the literature on shared GIS developments.

Consistency in Formats and Map Base

The obvious advantage from coordinated GIS and database activities lies in the fact that all parties in the shared arrangements studied used similar software and data formats. The fact that multiple agencies started to operate from a common base map was the major achievement. Our research confirms the experience expressed by many GIS users who argue that major benefits are reaped when the data are collected only once and used for multiple tasks. Clearly, coordinating and sharing databases improved operational efficiency. Interoperability was an inherent and enabling part of this achievement. The very nature of data interoperability is defined as the ability to access multiple, heterogeneous geoprocessing environments, either local or remote, by means of a single unchanging software interface (Buehler & McKee, 1996).

Enhanced Organizational Cooperation

Interorganizational GIS initiatives open up communication channels and allow an opportunity for close working relationships among employees at various organizational levels. In one case those intensified interactions were only temporary, but in four other cases they were long-standing. The communications at the highest administrative and decision-making levels were only occasional, but crucial for interorganizational projects to proceed. In one case the GIS database initiative helped revive a dialogue between a city and a county top leadership that was dormant (and unfriendly) for a long time, a fact that appeared in the headlines of the local newspapers. The communication intensity and frequency of contact were higher among the members down the organizational hierarchy. The joint GIS efforts especially served to strengthen networks among staff across functional departments. Those networks remained vital even when in two cases the formal relationships and coordinating structures were dismantled.

The group interaction helps enhance understanding of the technology. Cooperating users share ideas and jointly elevate their expertise, as previously asserted by Brown, O'Toole, and Brudney (1998). Research also confirms the previous claims that spatial data sharing leads to new relationships among parties (Calkins & Weatherbe, 1995). With intensified communications, the groups and individuals involved tend to redefine the nature of previous departmental rivalries, biases, and predispositions. The end result is that organizations move toward a closer, more collaborative economic relationship. Information technology and interorganizational systems play an enabling role in making this transition feasible (Clemons & Kenz, 1988; Clemons & Row, 1992; Reich & Huff, 1991).

GIS Diffusion to Small Jurisdictions

The findings of Nedovic-Budic and Pinto (2000) also bear out the argument that collaboration across multiple organizational boundaries has a positive impact on the diffusion of technology to other uninvolved groups. Once a dynamic of free data flow and open access has been institutionalized, smaller groups that did not attempt to develop geographic data before become (1) aware of software technology and (2) interested in using it. In the cases studied, there were several examples of small local government organizations that relied on the shared database developments to "jump-start" their own GIS. Once established, those local systems could contribute new data for their portion of the geography to the common areawide database, which was the ultimate goal of two-way interchanges. However, the usefulness of the regional development for local users depended on the quality of the database. In one case, the regional nature of the database, with its relatively low positional accuracy and currentness of parcel-level data, coupled with fees charged for database use, turned off several localities and led them to pursue their own basically parallel developments. The open and free access to the database was, on the other hand, very encouraging for several local users.

Transaction Costs of Data Sharing

Transaction costs are the costs of managing the interaction while keeping opportunistic behavior under control so that ongoing operations between the units can be sustained (Kumar & van Dissel, 1996). Interorganizational GIS and database activities carry with them the responsibility for active commitment to the arrangement from all involved parties. In many cases, member organizations are differentially positioned toward the joint arrangement and the transaction costs associated with its upkeep. Agreeing upon, managing, and achieving interorganizational GIS-related objectives requires investments of all organizational resources. The coordination in all cases took time, patience, tolerance, and energy that could have been in a short run more efficiently utilized at the organizational level. Initially, the belief that such an endeavor was worth it, and later the tangible demonstrations of the progress, kept the joint projects going.

Based on previous research findings, there is no question that interorganizational GIS activities are a resource drain. Unless there is a clear scope statement regarding the system's goals and tight project management, there is a strong tendency for schedule slippage, interdepartmental squabbling, and disagreement over resource commitments to complete the project. Critics of multiparty implementation efforts claim that partnerships are likely to experience circumscribed coordination, limited scale economies, frequent delays, and problematic outcomes (Brown,

Brudney, & O'Toole, 1998). According to Brown and colleagues (1998), with an average implementation time of 3–5 years before performance becomes routine, GIS partnerships must be more than spontaneous and ephemeral. Tight control and guidance of the implementation process is crucial for avoiding or minimizing the transaction costs.

Meeting the Agreements

Among the major difficulties cited by the interviewees were the problems of meeting agreements on equipment specifications, data standards, implementation time, and financial obligations. In the cases studied, as well as in the previous research, these difficulties frequently resulted in stalemates and deadlocks (Brown & Brudney, 1993). The most common reason for the unmet agreements was a lack of staff or financial resources to back the previously accepted responsibilities and procedures. Differences in needs also led to misunderstandings regarding the importance of particular requirements and standards.

Overall, interorganizational GIS and database activities were not easy to accomplish and had their costs. The overwhelming sense, however, derived from the analysis of cases, was that such efforts were worth the investment, and that the balance was on the side of benefits. The overall nature of data-sharing relationships, including the process flow of issues summarizing Nedovic-Budic and Pinto's (2000) research, are illustrated in Figure 9.3.

Clearly, "data sharing is easier to advocate than to practice" (Azad & Wiggins, 1995: 39). Managing change requires attention to many implementation issues identified in the cases studied, such as top management and administrative support for GIS and database development; secured long-term funding for the GIS and database projects; well-defined and focused project scope; timing of training, equipment, and system installation; the need to manage the users and their expectations about the degree, timing, and quality of data available; the importance of cross-organization communication to resolve disputes and misunderstandings; the need to demonstrate clear progress in order to allay political pressures; the necessity of managing personalities and private agendas; and existence of an identifiable champion or initiative leader. In his case studies of enterprise GIS solutions, Azad (1998) found management as the most important success factor. Well-managed projects were more likely to overcome the many difficulties associated with multiparticipant implementation process. While those difficulties were admitted across the cases

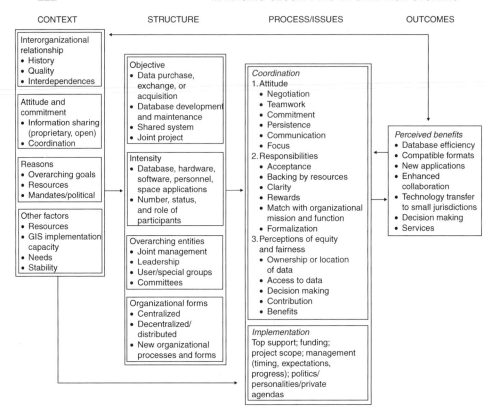

FIGURE 9.3. Summary of the case study results of Nedovic-Budic and Pinto (2000). From "Information Sharing in an Interorganizational GIS Environment" by Z. Nedovic-Budic and J. K. Pinto (2000), *Environment and Planning B: Planning and Design*, *27*, 455–474. Reprinted with permission from Pion Limited, London.

studied, the advantages and benefits of establishing interorganizational systems and databases were readily recognized too. The common base map was consistently mentioned as the most valuable product derived from coordinated GIS efforts.

Conclusions

In this chapter we have sought to create a context for providing a greater understanding of the problems associated with interagency information sharing and of some key factors in helping to mitigate these difficulties.

The study of information sharing within the context of GIS environments is still in its infancy. By making use of current organization theory and marketing channel research, we have attempted to create a framework for a better understanding of some of the causes of cooperation between organizations sharing geographic information. These streams of research have allowed us to make some supportable propositions concerning the set of causal flows between antecedents of information sharing and the likely results for an organization engaged in a cooperative venture. It is hoped that as a result of a better understanding of these linkages, managers and members of organizations tasked with the responsibility of developing cooperative relationships will be in a position to facilitate geographic information sharing to the advantage of all concerned organizations.

Chapter 10
Metadata for Geographic Information

It is readily apparent that the evolution of GIS has seen the technology transform itself from a specialized instrument for an elite group of experts into a useful tool for an aware and competent mass market. A good deal of the evolution, as you will have gathered by this time, is the outcome of standardization of the technology, the software, and the body of skills and knowledge associated with GIS. Not only have GIS technology, software, skills, and knowledge evolved, so too have data. Today, geographic data are available from a variety of public, private, and open sources, much of it at low or no cost to the data user. Equally important to the useability of these data are the fact that over the past decade metadata have become a routine part of the implementation of GIS.

This chapter discusses metadata. It begins with a description of metadata, and then offers a section detailing the reasons for metadata and the early calls for its development. The next section discusses the standardization of metadata, paying special attention to the roles of the National Spatial Data Infrastructure (NSDI), the Federal Geographic Data Commission (FGDC), and the International Standardization Organization (ISO). The chapter then describes the elements called for by metadata and how to provide them.

What Metadata Are and Why They Are Needed

Metadata are information about data (McDonnell & Kemp, 1995: 61). Within the context of GIS, metadata provide crucial information about the data that are the heart of GIS; this information in turn enables potential users to know if the data are acceptable for their specific application(s). Longley, Goodchild, Maguire, and Rhind (2001) liken metadata to the catalog of a library, in that metadata describe and organize data according to their attributes, including author, subject, date, projection, intended uses, limitations, and so on. Furthermore, metadata function as "the equivalent of documentation, cataloguing, handling instructions, and production control" (Goodchild & Longley, 1999: 574).

Like the traditional newspaper story, metadata provide answers to the basic journalistic questions what, who, why, how, where, and when. *What* geographic area do the data describe? *Who* gathered and processed the data? *Why* were these data chosen and put in the format used? *How* were the data collected and processed? *Where* were they organized? *When* were they first collected and when must they be modified (Indiana Geological Survey, 2003)?

A classic example making the case for the indispensability of metadata was related in Chapter 5 of this book. That chapter included a synopsis of Gersmehl's (1985) article "The Map, the Data, and the Innocent Bystander." Gersmehl described how the U.S. Department of Energy (USDOE) misinterpreted soil maps of histosols that he (Gersmehl) had previously produced and published. Because of its misinterpretation of Gersmehl's map, the USDOE incorrectly categorized and erroneously labeled a large class of soils as "peat," even though soils shown on much of the map were not of this highly specialized, energy-rich soil type. As a result, USDOE's maps of energy-producing peat were inaccurate, showing far more peatlands than actually existed. Gersmehl generously placed most of the responsibility for this misinterpretation at the cartographer's (his own) door. However, in the estimation of the authors of this book, Gersmehl goes too far in describing the map reader as an "innocent bystander."

It is just this sort of misunderstanding that Gersmehl describes, and the propagation thereof, that metadata are designed to avert. Generation and provision of metadata for digital cartographic products require the map*maker* to take responsibility for describing the capacities and limitations of the map. Reading, understanding, and acting within the limitations as described in the metadata are the responsibilities of the map *reader*—especially if that map reader plans to incorporate the map and/or other relevant geographic data into his or her own specialized map or GIS products.

In reality, metadata are equally helpful and necessary for both the mapmaker and the map user. Among other results, metadata may help the mapmaker who has provided proper metadata (including limitations) to avoid liability for the misuse of these maps and other GIS data sets by a map reader or map borrower who disregards the map's limitations, caveats, and restrictions. The responsibility of GIS managers, then, is both to create accurate metadata for their own GIS maps and products that may be used by subsequent users (Schurrman, 2004) and to review and assess the suitability of others' maps for specific applications based on the related metadata. As Goodchild and Longley note (1999: 573), "a dataset can be simultaneously the output of one person's science, and the input to another's."

Data quality lies at the heart of the need for metadata (Veregin, 1999). Goodchild and Longley (1999: 573–574) provide an example of how serious the lack of metadata can be for data quality.

> Suppose information on the geodetic datum underlying a particular dataset—potentially a very significant component of its metadata—were lost in transmission between source and user. . . . This loss of metadata, or specification of the data content, is equivalent in every respect to an actual loss of accuracy equal to the difference between the true datum and the datum assumed by the user, which can be several hundreds of metres.

Metadata may be viewed as a close relative of the health and safety warnings commonly found on consumer products (Heywood, Cornelius, & Carver, 1998: 153).

U.S. National Map Accuracy Standards: A Precursor to GIS Metadata

Although the push for metadata for GIS maps and data is a relatively recent phenomenon, the importance of a need for map accuracy has long been recognized. As early as 1941, the U.S. Bureau of the Budget developed National Map Accuracy Standards. The initial rules were issued in 1941, revised in 1943, and revised once again on June 17, 1947 (U.S. Bureau of the Budget, 1947; U.S. Geological Survey, 1999).

The National Map Accuracy Standards define standards in seven categories for the accuracy of published maps. The first two categories establish standards for horizontal and vertical accuracy. The third category describes how to test the accuracy of maps. The fourth category requires maps that meet the accuracy standards to note this in their legends.

The fifth requirement declares that maps that do not comply with the National Map Accuracy Standards must omit any mention of standard accuracy. The sixth says that maps that are enlargements of published maps should state this clearly in the map legend (e.g., "This map is an enlargement of a 1:24,000-scale published map"). The final requirement states that wherever feasible, maps "shall conform to latitude and longitude boundaries, being 15 minutes of latitude and longitude, or 7.5 minutes, or 3¾ minutes in size" in order to "facilitate the ready interchange and use of basic information" that may be shared among federal agencies (U.S. Bureau of the Budget, 1947). Certainly, the seeds for metadata were planted with the U.S. National Map Accuracy Standards.

The Standardization of Metadata

Calls for GIS metadata presented in a standardized format go back to the 1980s, that is, to the time when maps were becoming more readily available in digital form and GIS were becoming more readily available for a wider audience of users. With the proliferation of digital maps and data sets came the recognition that the rise in the number of digital maps and correspondingly increased amounts of digital data would make it increasingly easy to copy maps and other spatial data, which then could be shared by multiple parties. As copying and sharing maps and digital data became easier, the number of casual users was likely to rise accordingly.

Indeed, the push to share geographic information has been an important driving force behind the development and standardization of metadata (Beard & Buttenfield, 1999: 219; Guptill, 1999: 677; Schuurman, 2004: 64; Ventura, 1995: 173). Beard and Buttenfield (1999) specifically identify the development and proliferation of data libraries, including the growing use of the Internet for data distribution, as a major impetus to standardizing metadata. As Pinto and Onsrud (1997: 46) put it, "Information sharing among organizations is based on developing standardized or generalizable patterns of exchange." Without a doubt, standardization of metadata is a crucial part of this process.

Efforts to standardize metadata began in the 1990s. In the United States, Executive Order 12906 created the National Spatial Data Infrastructure (NSDI). The NSDI, in turn, created the Federal Geographic Data Committee (FGDC) whose mission is to coordinate the development of the National Spatial Data Infrastructure. The FGDC developed a data standard called "Content Standards for Digital Geospatial Metadata" (or

CSDGM), which were originally approved in 1994, and updated in 1998 (Indiana Geological Survey, 2003; Longley et al., 2001). Longley and colleagues (2001) emphasize that the CSDGM set the standard for *what* information should be included in basic metadata, hence the term "Content Standards." Significantly, the CSDGM do not provide guidelines about how metadata should be formatted or structured (Longley et al., 2001: 155).

Efforts to standardize metadata do not begin and end with FGDC. As Guptill (1999) notes, librarians have been using computers to create electronic data catalogs for years. The purpose of these efforts is to make it easier to search electronic data "primarily . . . to provide specifications for the exchange of bibliographic and related information between systems" (Guptill, 1999: 682). The standards for the metadata for these nonspatial data sets have been developed within the context of international efforts that have complemented U.S. standards. These standards were expressed in the Dublin Core (that's Dublin, Ohio, not Dublin, Ireland) with the goal of providing additional functionality in the standards, including the ability to enhance resource discovery and archive control.

While the Dublin Core and related metadata standards focus on more general data, the International Standards Organization (ISO) has been developing standards in the area of geographic information/geomatics (Salge, 1999). Indeed, as Salge notes, "The history of standardisation in the GI field extends over 25 years" (1999: 693), and exists in nearly 20 countries (Salge, 1999: 696). National standardization bodies from almost all the countries of the world comprise the ISO. The ISO Geographic Information Technical Committee 211 (ISO/TC 211) is the group designated to develop standards for world geographic information. ISO/TC211 included members from 24 countries and observers from 12 more countries, with specific cooperating agreements with a variety of professional organizations with expertise in geographic information. These organizations include the International Cartographic Association (ICA) and the International Society for Photogrammetry and Remote Sensing (ISPRS) (Salge, 1999). Metadata are among the standards developed by ISO/TC 211.

With multiple groups developing metadata standards, there would seem to be great potential for confusion. The good news for GIS managers is that there is considerable overlap among the various metadata systems that are currently available. Guptill (1999) also reminds us that metadata "do not require internal storage or display formats that are specific to individual systems," thus moderating the possible harm from choosing one set of metadata standards rather than another. Furthermore, there are ongoing efforts to make existing metadata standards interoperable, including efforts to create "crosswalks" between and among the existing standards by "identifying analogous or closely related items in each standard and using the same data value in each of those fields" (Guptill, 1999: 684). Another method is to develop a data mediator that will automatically generate data translations from descriptions of data in the source and receiver systems. Work to rationalize these systems is ongoing.

In the meantime, GIS managers need to review metadata from existing data sets to determine if these sets meet their needs. At the same time, GIS managers also have a responsibility to provide accurate metadata to accompany the data sets that they generate for their own use. Even if there is no immediate intention to share the data set with others, as the organization experiences staff turnovers, there will be a natural and growing distance between the people who created the data originally and those who use and modify it at later times. Weber (1945) would include metadata as part of the written "files" of a bureaucratic organization and insist on their critical importance.

Today, state-of-the-art GIS software includes templates for presenting metadata, easing some of the confusion regarding which standards to choose. Salge (1999) suggests that the geographic information industry (including software and data developers) will play a crucial role in the continuing evolution of metadata standards. Certainly the inclusion of metadata templates within GIS software lends credence to this prediction.

Elements of FGDC Metadata

As described above, several groups are developing metadata standards, but this should not cause the GIS manager to throw up his or her hands in frustration or use this ambiguity as an excuse to avoid developing metadata. Because, as Guptill noted, there is significant overlap in the metadata standards developed by the various groups, it is acceptable to choose from any of the metadata standards developed by reputable organizations, either in one's own country or by the ISO. Because the authors of this book are in the United States, we include the metadata categories identified in the Federal Geographic Data Committee (FGDC) standards, along with information about them. These categories are (1) identification information, (2) data quality information, (3) spatial data organization information, (4) spatial reference information, (5) entity and attribute information, (6) distribution information, (7) metadata reference information, (8) citation information, (9) time period information, and (10) contact information.

The FGDC suggests that providing a structure for geographic data is a key step in making data understandable to and therefore usable by consumers (Maitra & Anderson, FGDC, 2005). The FGDC website (*www.fgdc.gov*) provides valuable ongoing information about Content Standards for Digital Geospatial Metadata (CSDGM) on its website. The website is updated on a regular basis in order to keep pace with changes in technology and institutional mandates (FGDC, 2005). This chapter is intended merely as an introduction to metadata. GIS managers are advised to check in with the FGDC and other applicable websites periodically to ensure that they follow the most current metadata practices.

Identification Information

Identification information is basic information about the data set (FGDC, 2005). This is important for others who may wish to use the data at a later time, providing them with a general sense of the usefulness and appropriateness of the data for their use based on the similarity of their purposes and data coverage areas with those of the originators of the data set. Identification information should describe the spatial data set by beginning with a succinct abstract of the nature of the information. It is important in this section to detail the purpose for which the data have been collected, along with the time period covered by the content, and the geographic coverage of the data set.

The identification information must include some additional information that is crucial, namely, access and use constraints. Access and use constraints are restrictions and legal prerequisites for accessing the data set. Most often, access constraints are designed to protect the privacy of individuals included in the data set, or to protect intellectual property rights associated with the data. Identification information also includes point of contact information that provides the name and other contact information for an individual or organization with knowledge about the data set.

If you are a GIS manager reviewing data sets developed by others, identification information will provide you with the basic description of the data that will enable you to determine if the data are appropriate for your application. Were the data collected for the same purpose or a similar purpose to those of your organization? Do the data cover the specific time period that you need, or at least a time period that complements and enhances your own needs? And do the data cover the same geographic area as the area for which you are building a GIS? When in doubt, get in touch with the individual or organization named as a knowledgeable contact for the data set.

Data Quality Information

Information about data quality requires a general assessment of the quality of the data set (FGDC, 2005). The Spatial Data Transfer Standard (SDTS) provides recommendations on the specific information about data quality that should accompany data sets, along with specific tests that can be applied to the data to determine its quality. Positional accuracy, attribute accuracy, and lineage are all included in the data quality information.

Positional accuracy requires as a first step an estimate of the accuracy of the horizontal positions of the spatial objects in the form of a Horizontal Positional Accuracy Report. This report is "an explanation of the accuracy of the horizontal coordinate measurements and a description of the tests used" (FGDC, 2005). Similarly, an estimate of the vertical positional accuracy is also required, along with a Vertical Positional Accuracy Report, which is analogous to the report for horizontal accuracy (FGDC, 2005).

Attribute accuracy is an assessment of the accuracy of the identification of entities and the assignment of attribute values in the data set (FGDC, 2005). Are the objects in the data set accurately described and categorized? Assessments of attribute accuracy must include several important elements, including their positional accuracy (Are they located where they actually exist?), along with reports on logical consistency and the completeness of the attribute data (FGDC, 2005).

Lineage is a crucial element in GIS metadata because it gets to the heart of the problem of propagation of error. Heuvelink (1999: 207) succinctly describes this problem:

> The data stored in a GIS have been collected in the field, have been classified, generalised, interpreted or estimated intuitively, and in all these cases errors are introduced. Errors also derive from measurement errors, from spatial and temporal variation, and from mistakes in data entry. Consequently, errors are propagated or even amplified by GIS operations.

Lineage is the means by which developers of GIS data report the steps in the process from original data collection to presentation of the data in its current form. "Lineage" is analogous to the antique community's concept of "provenance" (made familiar to a more general audience by the PBS series *Antiques Roadshow*).

The FGDC defines lineage as "information about the events, parameters, and source data which constructed the data set, and information about the responsible parties" (FGDC, 1994). Properly presented, the lineage of a data set includes each step along the path that brought the data to its current status, starting with its source and including each step of processing that transformed the data. Furthermore, the FGDC standards call for a contact for the source data as well as a contact for each processing step.

As a manager integrating data from a previous user, you need to review and understand the nature of the original data along with the processes that transformed the data. If there is any uncertainty, you would be wise to call the contact people to clarify any questions. As a manager preparing geographic data that may later be shared with others, you need to provide clear descriptions of any transformations that have altered the data in any way.

Finally, the data quality information must also include information on cloud cover for any remotely sensed and other data that might be affected by this phenomenon.

Spatial Data Organization Information

Spatial data organization information refers to "the mechanism used to represent spatial information in the data set" (FGDC, 2005). This includes both indirect spatial reference information, such as the names of specific types of geographic features or addressing schemes that provide locations to objects in the data set, and direct spatial referencing, which is the system of representing raster and vector data in the data set.

Spatial Reference Information

This information describes the frame of reference (including the projection) for, and the means to encode, coordinates in the data set (FGDC, 2005). This information is crucial because spatially referenced information must always use the same reference points in order to be combined together as separate layers in a GIS. The projection system used in representing the data and the scale, as well as the resolution of the data, are needed in this section.

The FGDC guidelines for spatial reference information are extensive. It is useful to know, however, that today's GIS software often have the capability to identify the spatial reference information for most data sets, and to transform data sets from one spatial reference system to another. Moreover, it is common for GIS projects to incorporate base maps that are readily available from government or other sources, especially when those maps are offered at low or no cost and are of known and acceptable quality. These factors do not, however, relieve the GIS manager of his or her responsibility for understanding the information presented in this part of the data set and providing it for future users of the information.

Entity and Attribute Information

A key element of any GIS are the entities and attributes included within the GIS. Veregin (1999) includes the elements of entities as attributes in his discussion of "theme" as an important component of GIS. According to Veregin (178):

> Geographical phenomena are not really about space, but about theme. We can view space (or more precisely space-time) as a framework on which theme is measured. It is true that without space there is nothing geographical about the data, but on the other hand without them there is only geometry.

Theme, essentially, is why humans have created maps for millennia, and why today we build GIS. Entities and attributes constitute the theme, or the phenomena whose locations we wish to map for some purpose. Whether we are talking about Mercator's navigation charts or the Centers for Disease Control's (CDC) maps of cancer rates in the United States, or the McDonald's Corporation's maps of its store locations, we are talking about theme or, in FGDC parlance, entities and attributes.

An *entity* is the specific phenomenon located on the map, such as the location of the U.S. Environmental Protection Agency's Superfund sites. The *attributes* are the defined characteristics of such sites, such as the severity of the contamination and the state of the cleanup. The quality of entity and attribute information depends on several factors, including the due diligence with which the original data are collected, the care in entering these data into the GIS, and the degree to which the passing of time renders the data obsolete.

Veregin (1999) identifies temporal accuracy as a critical element of assessing the data quality of GIS themes. This idea is echoed in the FGDC guidelines for entity and attribute information. In addition to providing an accurate definition of entities and attributes, ensuring that each of these is regulated by a valid range of values (either quantitative or qualitative), the FGDC also calls for the dates for which the data associated with entities and attributes are current. Different types of entities are subject to obsolescence in different time frames. For example, the locations of continents is reasonably stable from one millennium to the next. By contrast, the locations of new homes in a hot housing market can expand dramatically from one year to the next.

The careful GIS manager will carefully assess the temporal aspect of entities and attributes before importing such data into his or her GIS. Similarly, a responsible manager will be equally careful to describe the temporal framework of any data that may be exported from his or her GIS, and perhaps even provide an expiration date (or at least a strong caution) on entity and attribute information that is subject to quick obsolescence.

Distribution Information

This is information about the distributor of the data set and options for obtaining the data set (FGDC, 2005). It is apparent that this section is intended primarily for organizations who expect to make their databases available on a relatively regular basis. The metadata to be included in this section contain technical information about the data itself, along with protocols for data transfer.

One item of particular interest in this metadata category is "Distribution Liability," which is a statement of the liability that the distributor assumes. Liability for shared data is discussed at length in Chapter 13.

Metadata Reference Information

This is information about the currentness of the metadata information and the responsible party. It begins with information about the date the metadata were created or updated, along with the date of any reviews of the metadata. This section requires information on the metadata standard used, and the date by which the metadata entry should be reviewed.

This section includes information regarding use constraints, which detail any restrictions or legal requirements for using the metadata after access is granted. In addition, this section includes information about handling restrictions "imposed on the metadata because of national security, privacy, or other concerns" (FGDC, 2005).

Citation Information

Citation information is the recommended format of the reference for a data set. Citation information begins with providing the name of the organization or individual who constructed the data set. If editors or compilers have worked with the data, they should also be identified, with their roles ("ed." or "comp.") indicated. The publication date is needed. Time of day when the data set was published or made available may also be included. Citation information must also include the title by which the data set is known, along with the version or edition of the dataset.

Other information to be included are series information (if the data set is part of a larger series), online linkage (the online computer resource that contains the data set), and a larger work citation (the information identifying a larger work in which the data set is included).

Time Period Information

Veregin (1999: 178) observes that "time is not given sufficient attention" in GIS. He goes on to say:

> Although poorly accommodated in conventional geospatial data models, time is critical to an understanding of geographical phenomena, not as entities that exist in some location, but as events that appear and disappear in space and time.

The FGDC language regarding time period information reflects Veregin's view. Time period information is defined by the FGDC as "the date and time of an event." This category allows the metadata developer to include information on a single date and time, multiple dates and times, or a range of dates and times. And as the directions for this section point out, the ways to describe temporal information indicated in this section are applicable in all other parts of the metadata standard. This section is always used in conjunction with other categories of metadata, never alone.

Contact Information

This is critical information, for it provides the identity of the person(s) or organization(s) that prepared the data set. It includes names, addresses, telephone and fax numbers, e-mail addresses, and hours of service, along with any other specific contact information needed to facilitate communication between the developer of the data set and any subsequent consumers of the data.

"Don't Duck the Metadata"

The FGDC has a public relations campaign to encourage GIS data developers and consumers to provide complete metadata; it uses the catchphrase "Don't duck the metadata." A number of organizations offer short courses on metadata that are extremely useful for those who must either interpret metadata or who must create metadata as they develop their own GIS projects. Colleges and universities, state geological surveys, the FGDC, and other organizations offer such training, often at no cost. It is worthwhile to explore these avenues and take advantage of these resources.

While much of the impetus for metadata comes from the increasing propensity to share databases across organizational boundaries, all GIS databases need metadata. Even if an organization knows (or believes) that it will never share its data outside its institutional boundaries, metadata should be viewed as part of the organization's permanent records, its files, or its institutional memory, as Weber (1945) would describe them. Employees may come and go and with them their personal knowledge of the data and its quirks. Only by having—and maintaining—metadata, will the organization maintain an institutional memory of the data that undergirds its GIS.

Data is at once a significant expense associated with GIS and the very foundation upon which it is built. Metadata is the way we know our data. Without this knowledge, the data themselves are of questionable value, and so too is the GIS. Organizations must be as conscientious about metadata as they are about the data themselves.

Chapter 11

Policy Conflicts and the Role of GIS
PUBLIC PARTICIPATION AND GIS

In the first edition of this book, we discussed an early claim by GIS advocates that the technology could help to minimize conflicts over land use by providing more and better information about the subject of the conflict. At the time, we pointed out that this claim overlooks an important source of conflict: the underlying value differences represented by conflicting parties. In the first edition, our chapter on the topic explored the role of GIS in the public policy sphere, and suggested that GIS would tend initially to increase rather than to decrease conflict since geographic information and analyses made possible by GIS can be used selectively by conflicting parties to support their positions. We further asserted that this conflict can— and should—be viewed as a positive feature in a democracy since it represents an open dialogue concerning differences of opinion that must be fully explored as a precondition for acceptable public policy resolutions.

In fact, our prediction has become reality with the evolution of the public participation GIS or PPGIS movement. PPGIS is the shorthand way to refer to the use of GIS by nongovernmental, nonbusiness organizations to make their voices heard and, if possible, obeyed, in spatial conflicts in which they have a stake. While inspired groups were using GIS earlier, the PPGIS movement really took flight with an initiative and specialist meeting of the Varenius Project of the National Center for Geographic Information in 1998. Like its predecessor in the first edition, this chapter discusses the original claim that GIS would help minimize spatial conflict, and follows up with a discussion of the evolving PPGIS movement.

In addition to their tremendous potential within private organizations, GIS have become an increasingly important tool in the public sector, and more recently among nongovernmental organizations (NGOs). Through most of the history of the technology, the primary orientation of GIS to date has been toward information and infrastructure management rather than spatial and policy analysis (Anselin, 2002; Goodchild & Getis, 1991). Today, GIS is growing in importance as a tool for analysis and public policy development, including such tasks as land use analysis and political redistricting (Fotheringham & Rogerson, 1995; Sawicki & Peterman, 2002). It has been suggested that GIS may play a role in minimizing conflicts among competing interests regarding land use issues by making available more and better information. "Better information" is usually construed to mean information that is more current and accurate. It is hypothesized that this information will improve analysis and facilitate agreement among the competing parties. This suggestion appears to be somewhat naive since it overlooks the underlying value conflicts that precipitated the initial conflict in the first place.

Rather than minimizing conflict, we originally hypothesized that the ready access to information made possible by the proliferation of GIS will lead to *increased*, not *decreased*, conflict in the short run as a greater number of diverse interests harness this powerful tool in support of their objectives, and that eventually the conflict will level off, although at a higher level than previously existed. While we have not done a study to assess the level of conflict before and after the proliferation of GIS, anecdotal evidence undoubtedly shows that savvy organizations are using GIS as a tool to make their voices heard in public policy debates.

The logic behind our suggestion that GIS has the potential to increase conflict lies in research that identifies two sources of conflict: disagreement on facts (cognitive conflict) and disagreement regarding values (interest conflict). While GIS can have an impact on facts in a particular conflict, there is no reason to expect that it will do anything to mesh competing values. Value conflict, therefore, will remain, regardless of the amount of information gathered to resolve it. At the same time, more information increases the number and variety of "facts" that can become the basis for further conflict. In short, GIS enables groups to provide alternative data and explanations while they propose alternative solutions to conflicts.

This chapter begins by providing a discussion of the theoretical underpinnings of cognitive versus interest conflict. It follows with a discussion about how cognitive and interest conflict can respond to the addition of a GIS as an analytical tool. We then provide examples of groups using GIS to push their program in public policy debates. The chapter concludes by discussing the value of conflict in a democracy and the role of GIS in that conflict.

Cognitive versus Interest Conflict

Among the potential benefits of implementing GIS, the question has been raised: Can GIS minimize conflicts regarding land use? The belief that GIS will help minimize conflict begins with the premise that GIS makes more readily available and accessible greater quantities of data and information within the framework of a computerized package for analysis of the data. This belief rests on two basic assumptions: (1) that more information is necessarily better; and (2) that all participants in the conflict agree on the validity of both the data and the decision models used within the framework of the GIS. In short, this view assumes that there is an objective reality on which all parties can agree. It is our contention that these assumptions are inaccurate, a contention that is consistent with recent literature on conflicts in policymaking as well as with earlier theoretical work of Weber (1968a) and Habermas (1981). Evidence drawn from the public participation GIS literature also supports our earlier hypothesis.

Weber challenges the assumption that there is an objective reality on which people can agree, basing his argument on the existence of differing (and often competing) values that people hold near and dear. According to Weber, debates that have values at their core represent a source of irreconcilable conflict. No amount of rational appeal is guaranteed to sway an individual from his or her values. In fact, argumentation based on facts and designed to change values—substantive rationality—is often futile. In Weber's view, "Scientific pleading is meaningless in principle because the various value spheres of the world stand in irreconcilable conflict with each other" (Gerth & Mills, 1976: 147).

Weber describes argumentation based on values as being "ends rational." By ends-rational action, Weber refers to the goal choices of individuals, closely intertwined with individually held values. Ends-rational actions hold intrinsic value for the actor, regardless of the outcome, and are (obviously) ends in and of themselves. For example, animal rights activists, such as PETA (People for the Ethical Treatment of Animals) who demonstrate at fur fashion shows and intimidate potential fur buyers and fur sellers undertake such activities as a personal mission. While they may make a few people change their minds, their actions do little to sway public opinion or to outlaw the sale of furs. Indeed, the aggressive tactics of groups like PETA and ALF (Animal Liberation Front) do more to hurt their cause in the court of public opinion than members of these groups may realize. Still, in their minds, these activists are doing the right thing, regardless of the long-term outcome. Winning the battles appears to be enough, even if they ultimately lose the war (or so it seems). The protests themselves often have intrinsic value for the protesters.

In contrast, Weber discusses the notion of formal rationality, which is "means rational." By "means-rational action," Weber refers to the strategy an individual follows to achieve his or her goal, a strategy that may include short-term losses. The means-rational person chooses a strategy that has the greatest chance of achieving his or her ultimate goal, and is willing to forgo short-term gains for long-term success. People and groups choosing means-rational action do not mind losing a battle if it ultimately means winning the war. Since values are not constantly at the forefront for the means-rational actor, discussion and persuasion based on information and logic are likely to be more successful in achieving resolution in this situation. This person or group might also be more open to compromise.

Again, to offer an example, consider the case of the passage of a pre-scription drug bill in the U.S. Congress in the summer of 2003. The bill, originally sponsored by Republicans, eventually gained the support of long-time nationalized insurance advocate Senator Ted Kennedy of Mas-sachusetts. The bill included a much-debated involvement by private insurance companies. Purists and many long-time national insurance advocates saw the bill as seriously flawed, but they were willing to compro-mise, believing that this would be a first step toward a better policy later (or the camel's nose under the tent, as seasoned budgeters would describe it). This exemplifies means-rational action.

Brubaker (1984: 36), critiquing Weber, comments that "formal ratio-nality is a matter of fact, substantive rationality a matter of value." Policy decisions are almost always value-laden; hence the use of a GIS in land use analysis and policy development would do nothing to minimize the diver-gence of values held by the participants in the policy process, but instead would only insert a new tool into that process. At most, a GIS would make more information available more rapidly. The addition of more facts at a more rapid pace may, in fact, only lead to increased conflict, at least in the short run.

In contrast to Weber, Habermas emphasizes the existence of an objective reality and is less concerned about values. In conflicts re-garding this objective reality, the more rational argument—that which raises greater validity claims—should hold sway (Habermas, 1985: 99). Habermas emphasizes the role of issue definition in disagreements among interested parties:

> A definition of the situation by another party that prima facie diverges from one's own presents a problem of a peculiar sort; for in cooperative processes of interpretation no participant has a monopoly on correct interpretation. For both parties the interpretive task consists in incorporating the other's interpretation of the situation into one's own in such a way that in the revised version "his" external world and "my" external world can—against the background of "our" lifeworld—be relativized in relation to "the" world, and the divergent situation definitions can be brought to coincide sufficiently. (100)

Rational communicative action would be the means to accomplish this task and would ultimately result in agreement. This agreement will then form the foundation for later argument regarding the resolution of the issue itself. According to Habermas, the participants will ultimately settle their disagreement by presenting their arguments to each other, with the more rational argument winning.

Habermas stresses the importance of developing a framework of social norms among the parties engaged in discussion. His concept of communicative action presumes an objective reality about which meaningful argument can occur, permitting the participants to reach a logic-induced understanding. Weber's concept of substantive rationality is directly at odds with Habermas's concept of rationality as defined by his theory of communicative action. First, Habermas's theory assumes that reaching understanding is an important goal of human interaction. Weber (1968b), on the other hand, recognizes that "rational exchange is only possible when both parties expect to profit from it or when one is under compulsion because of his own need or the other's economic power" (73).

Furthermore, he emphasizes that "not ideas, but material and ideal interests, directly govern men's conduct" (Gerth & Mills, 1976: 280). In human terms, Weber's theory holds that most people engage in conflict for personal as well as professional reasons and that self-interest is often the driving force behind these conflicts and their possible resolution. Habermas seems implicitly to assume that when all the facts are in, the logical decision will hold sway as those in the wrong see the error in their viewpoints and capitulate to some objective "truth." Weber takes a more realistic (socially driven) approach that suggests that people often invest far too much emotional capital into their positions to surrender grace-

fully in the face of opposing argument, confrontation, or even direct logical refutation of their views.

Anyone who has ever watched a *"Crossfire* kind of new reality show,"
especially popular on cable TV, has seen for him- or herself how value-laden
discussions end up producing plenty of heat, but little light—and virtually
no mutual understanding, compromise, or resolution. If any format provides overwhelming evidence to support Weber's theory that deeply held
values hinder conflict resolution, then what passes for political discourse
on cable TV at the beginning of the second millennium is it.

Against this theoretical backdrop, more recent studies of land-use
decision making pick up the theme of ideas (facts) versus interests (values), and suggest that science and technology alone will not minimize
conflicts. In their analysis of land-use studies using multiobjective programming models, Wang and Stough (1986) discovered that most such
models fail to consider the cognitive processes of decision makers, and
note the importance of cognitive conflict within the policymaking process. According to Wang and Stough, "Cognitive conflict exists when individuals base their decisions on different facts or on the same facts construed differently" (107). By contrast, interest conflict occurs when
individuals have different values or desire different outcomes. Belief in
the ability of GIS to minimize conflict rests on an implicit assumption
that the participants in the decision-making process agree on the relative
importance of the facts (data), that all participants construe those facts
identically, and that all are motivated to compromise.

Other authors (Brill, Flach, Hopkins, & Ranjithan, 1990; Stough &
Whittington, 1985) have responded to the problem of interest conflicts
with the development of multiobjective models that produce an array of
resolutions (as opposed to a single "best" resolution) to land-use conflicts.
This strategy may prove more hopeful as a conflict resolution technique,
and offer a constructive role for GIS. GIS is uniquely adapted to develop
and graphically show alternative scenarios that could facilitate compromise among competing interests. The potential is there, if the participants
are motivated and willing to submit to rational argumentation.

A Model of Conflict

The notion of two very different types of conflict (based on facts or based
on values), then, is well accepted. If we accept this idea, we can easily envision a situation wherein any given debate over land use will have two components, interest (value-based) conflict and cognitive (fact-based) conflict.

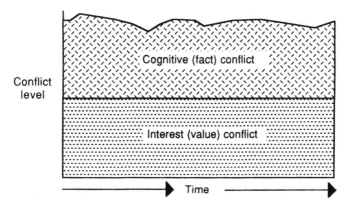

FIGURE 11.1. Routine conflict level before introduction of GIS. From Obermeyer (1994). Copyright 1994 by URISA. Reprinted by permission.

Since many such debates are long-lasting, we can further imagine an ongoing conflict over a period of time, with interest (value) conflict remaining more stable than cognitive (fact) conflict. Figure 11.1 illustrates this model.

An Example of Conflict

A real-life example that illustrates cognitive and interest conflict may be drawn from the case of a nuclear generating facility that was proposed for siting immediately adjacent to the Indiana Dunes National Lakeshore, in northwestern Indiana (Obermeyer, 1990a). First proposed in 1972 by the Northern Indiana Public Service Company (NIPSCO), the Bailly plant, as it was called, was designed to provide energy to homes, business, and industry (including the once-thriving steel industry) in the region.

A group of homeowners in the nearby residential areas that have been described as "among America's finest exclusive communities" (Mayer, 1964) learned of the proposed facility and began a successful decade-long legal war to derail the project. With the advantage of higher-than-average education and wealth, members of the group possessed both the knowledge and the economic resources to wage an extensive legal war against a major public utility. These residents united and called themselves the "Save the Dunes Council" (Obermeyer, 1990: 80).

Throughout its history, competing interests have consistently been at the heart of controversy surrounding the Dunes. Certainly, economic considerations played a key role in opposition to the proposed Bailly nuclear plant. However, Engel (1983) claims that more personalized and even spiritual attachments to the Dunes have been an important element of all controversies engulfing these "Sacred Sands."

At the heart of the conflict surrounding the siting of the Bailly nuclear generating facility was a difference in values or interest between the public utility and the Save the Dunes Council and its allies (also called "joint intervenors"). NIPSCO's interest lay in its ability to provide an adequate supply of electricity for its residential and commercial customers. The council's interest lay in its desire to protect and preserve the residential communities in which the members lived from the intrusion of such a large and potentially harmful facility and its consequences, both intended and unintended. Later, environmental groups and two local labor unions joined with the council to promote their own interests, which were different from those of the Save the Dunes Council, but led them to oppose the siting of Bailly. The council expressed its concern about the Dunes as an effort to protect an important environmental area. However, it is hard to imagine that the members were not at least equally concerned about the effects—both real and potential—of the proposed facility on their homes and property values.

Not surprisingly, given the demands of our legal system, both the council (and its allies) and the utility (NIPSCO) used facts to support their positions in challenges, counterchallenges, and rebuttals brought before the Nuclear Regulatory Commission (NRC) and the courts. For example, the Save the Dunes Council and its allies challenged NIPSCO's choice of a site for the Bailly facility, charging that the NRC had violated its own regulations regarding the acceptable distance between a nuclear generator and population centers. The joint intervenors argued that the NRC should have calculated the distance from the Bailly site to the border (not the center) of the nearest municipality.

Failing in their challenges in administrative hearings and appeals within the NRC, the Save the Dunes Council and its allies took their case to the U.S. Court of Appeals, which found in their favor. NIPSCO's appeal of that decision sent the case to the U.S. Supreme Court, which reversed the Court of Appeals' decision, and held that the NRC was the official expert on nuclear energy, and therefore the NRC alone was entitled to decide on an appropriate method for calculating the distance between the facility and the nearest population center.

Later on, two local chapters of the United Steelworkers joined the Save the Dunes Council and allies in the council's concern about the method for calculating the distance between the proposed nuclear facility and population centers. The Steelworkers' concern was that NIPSCO's method for calculating the distance between the facility and the steel factories exaggerated the distance between their workplace and a potentially deadly neighbor, in case of a nuclear accident. These workers worried about their ability to evacuate their nearby factories in case of a nuclear accident at the proposed Bailly facility. And while they agreed with the Save the Dunes Council that NIPSCO's distance calculations were flawed, their specific interest was neither in protecting the exclusive residential communities of the region, nor in saving the Dunes for conservation's sake. Rather, they were concerned about their own personal safety in the case of a nuclear emergency. Plausibly, had NIPSCO succeeded in delivering an acceptable evacuation plan using more realistic distances, the Steelworkers would have been satisfied. The differences between the Steelworkers and the Save the Dunes Council suggest that the two groups agreed at the level of cognitive conflict, but not at the level of value conflict. Still, to the extent that their agreement on the unacceptability of the proposed Bailly site could enlarge and strengthen the opposition against a common opponent, and perhaps increase their chances of defeating it, the unity of these disparate groups was assured. The joint intervenors (especially the environmental groups) and NIPSCO also disagreed about other "facts" in the case, too numerous to mention here. These facts formed the basis of a number of different administrative and legal actions against NIPSCO.

While we can separate the cognitive conflict from the interest conflict in the Bailly case on an intellectual basis, it is also clear that the two types of conflict are closely interrelated. More significantly, it appears that the cognitive conflict is driven by the existence of interest conflict in this case. Bluntly put, the joint intervenors used cognitive, fact-based conflict as a way to promote their own interest and values, which they believed were best served by the demise of the proposed Bailly nuclear generator. Again and again, over the course of nearly 10 years, the Save the Dunes Council and allies raised challenges based on facts in the case. Sometimes they won; more often they lost the battles. However, their persistence paid off in the end.

A decade after NIPSCO had first proposed the Bailly nuclear facility, it withdrew its proposal. By that time, NIPSCO had succeeded only in digging a very large hole, at the cost of some 200 million dollars. Ultimately, the direct and indirect costs of addressing the administrative and legal challenges of the joint intervenors became overwhelming, and NIPSCO was forced to back down. Later, the utility did build another electrical generating facility, a coal-fired plant in Wheatfield, Indiana. Construction of that facility went on without opposition, very likely because the proposed site was in a sparsely inhabited area.

A Hypothesis about GIS and Conflict

As the Bailly case study suggests, in practice as well as in theory cognitive conflict and interest conflict are separate intellectual concepts, even though they are often intertwined, practically speaking. Furthermore, this case suggests that issues raised at the level of cognitive (fact) conflict may be used to promote specific interests (values). We must keep in mind that the primary forum in which such conflicts are played out is the U.S. legal system, which is an adversarial system that relies heavily on facts. The system is inherently conflictual.

Among other things, the Bailly case demonstrates the value of using specific cognitive issues to challenge and rebut opponents in the courts. Although the Save the Dunes Council and allies almost certainly preferred winning to losing, and a big win, either in the administrative legal environment of the NRC or in the U.S. judicial system would have meant the immediate and final demise of the Bailly facility, just staying in the game advanced the council's cause. Ultimately, staying in the game enabled them to win what turned out to be a game of attrition. Information enabled them to keep the game going.

In the original edition of this book, we expressed our belief that "it is not difficult to imagine the value of a geographic information system as a means to produce more information that may be used to support specific positions on a given land-use issue, especially in a public setting." We pointed to the Freedom of Information Act, which helps to ensure open access to nearly all government-owned information. In addition, we mentioned additional efforts by the federal government to improve public access to certain types of base maps and other geographic information (e.g., the TIGER files). The result, we suggested, is that much of the georeferenced data that can be used in implementing a GIS is readily available in a GIS-compatible format.

In addition, we noted, the price of GIS hardware and software is decreasing in real terms. As computers themselves have become accepted as commodities, manufacturers of the equipment face pressure to reduce prices in order to remain competitive. Just as important, technological improvements have increased the power of both hardware and software.

The combined result of these trends is that GISs are becoming more affordable and more prevalent. Furthermore, as GIS become more widely used, more people and interest groups have begun to harness this technology to help them build and support their arguments on land-use issues. Because of the ability of GISs to make more information available more quickly, it logically follows that groups that adopt GIS are able to develop more and perhaps stronger facts and arguments to support their positions.

The likely result will be more conflict, not less, as more groups harness GIS to help them build support for their positions on particular land-use issues. In addition, that increased conflict is likely to be cognitive, rather than interest, conflict. Interest conflict should remain steady. Figure 11.2 illustrates the hypothesized change in conflict over time with the introduction of GIS.

A logical question would be to ask how conflict is expected to increase as the result of enhanced access to information. After all, if two (or more) parties are presented with the same objective data, should not such information resolve misunderstanding and lead to mutually acceptable resolution? If all conflict were simply cognitive (i.e., based on factual disagreement), the answer may perhaps be yes. However, value conflict leads to a level of emotion that is not easily influenced by the mere pre-

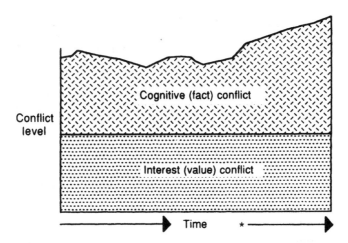

FIGURE 11.2. Hypothetical conflict level after introduction of GIS. Asterisk indicates time of GIS introduction. From Obermeyer (1994). Copyright 1994 by URISA. Reprinted by permission.

sentation of objective data. Furthermore, we must realize that conflicting parties often interpret the same data in very different ways. An example is the different "spins" put on the results of research into the dangers of cigarette smoking when interpreted by the Tobacco Institute versus the U.S. Surgeon General's office (before their recent epiphany and major cash settlements). All forms of data are only as useful as their interpretation. It is this difficulty with objective interpretation, however, that lies at the heart of conflicts resulting from increased access to information.

Public Participation GIS: Good News for a Democracy

In the intervening years since the publication of the first edition of this book, there has been a tremendous growth in the use of GIS by organizations of all stripes—private, public, and nongovernmental. Of greatest relevance to our hypothesis regarding spatial conflict is the growing use of GIS and spatial data by nongovernmental and grassroots organizations to advance their policy goals. Sawicki and Peterman (2002) identify over 60 NGOs using GIS as a tool to influence policy, and the number is growing. The use of GIS by NGOs involves a variety of settings including urban, rural, and environmental. Such use is not limited to the United States and other developed places, but exists in less developed countries and regions as well (Bond, 2002; Kyem & Kwaka, 2002).

A prime example of an urban nongovernmental organization using GIS is the Metropolitan Area Research Corporation in Minneapolis, Minnesota. This nonprofit organization uses GIS as a tool in its study of "the growing social and economic disparity and inefficient growth patterns in metropolitan areas" (National Research Council [NRC], 2003: 116). A specific goal of this organization is to "assist individuals and groups in promoting greater equity" (NRC, 2003: 116). Also in Minneapolis, Elwood (2002) reports on the use of GIS by an NGO in one of the city's neighborhoods, Powderhorn Park. In cities like Philadelphia (Casey & Pederson, 2002), San Francisco (Parker & Pascal, 2002), Atlanta (Sawicki & Burke, 2002), Chicago (Al-Kodmany, 2002), and others, GIS is also used as a tool by NGOs to influence public policy.

It is not only urban NGOs that have adopted GIS in order to pursue their policy goals. Around the world, environmental organizations have adopted GIS to advance their policy objectives. Sieber's (2002) research and experience using GIS within the environmental context demon-

strated that environmental groups use GIS "with passion and for advocacy" (169). Tulloch (2002) echoes this finding, noting that "environmental groups are . . . increasingly empowered users of geospatial
technologies" (192). In places as far-flung as Newfoundland (MacNab,
2002), Nepal (Jordan, 2002), South Africa (Harris & Weiner, 2002), and
Australia (Walker et al., 2002), NGOs have used GIS to advance their policy objectives.

One theme that is often repeated within the PPGIS movement is the
importance of data (facts) in advancing policy objectives, lending support
to the hypothesis that the use of GIS will tend to increase fact-based or
cognitive conflict. For example, Weiner, Warner, Harris, and Levin
(1995) identified significant discrepancies in lands identified as suitable
for agriculture by the apartheid government in Kiepersol, South Africa, as
compared with those so defined by black South Africans in the area. Government officials, using definitions that included an implicit assumption
that modern, high-tech agricultural machinery would be used, excluded
from its map of areas suitable for agriculture a substantial amount of land
that black South Africans actively cultivated. The reason for the discrepancy became apparent: black South African farmers, lacking the financial
ability to pay for modern farm equipment, used traditional tools, techniques, and practices, such as beasts of burden, raw human power, and
hand tools. This "outdated" system enabled traditional farmers to cultivate land that might have been too steep or otherwise too rugged to grow
crops in places where large modern farm equipment was impractical
(Weiner et al., 1995). Similarly, in the United States, many Amish farmers
have recently settled in the hilly terrain of rural Indiana counties that
"modern" farmers avoid as wasteland. Here we see examples of "the same
facts construed differently" (Wang & Stough, 1986: 107).

Community leaders and GIS users are well aware of how valuable spatial data can be in empowering people to make their case in the public
policy arena. In a National Research Council report, Stella Adams, a
leader of the North Carolina Affordable Housing Center, makes this point
passionately: "Having access to GIS empowered me to be able to do things
I couldn't. [People} need the proof to show to elected officials. A map
gives them legitimacy. Then you're paid attention to" (National Research
Council, National Academy of Sciences, 2003: 53).

While spatial data are important to NGOs, so too is the value of GIS
as a tool to visualize that data. It is useful to keep in mind, however, that
results of visualization using GIS may not always produce the desired
effect. For example, Karl Kim (1998) reported on his own experience in

"Using GIS Technologies to Empower Community Based Organizations in Hawaii" to challenge a utility company seeking to place a high-voltage electrical powerline along an unspoiled ridge jutting into the Pacific. Environmentalists enlisted Professor Kim's assistance to use the capabilities of the GIS to construct a virtual, three-dimensional image showing what the new towers would look like in order to demonstrate how visually damaging they would be to this beautiful coastal setting. These images would, the environmental group believed, sway the planning board and public opinion irrevocably against the powerline. As it turned out, while the images did not delight the utility company, neither did they satisfy the environmentalists. The images, showing correctly scaled towers, were not nearly as disfiguring as the environmentalists had imagined (or perhaps hoped). This GIS effort did not resolve the issue either way, but it did enable both sides to get a preview of the powerlines, which had the benefit of bringing useful information to the discussion.

While the first reaction to the hypothesis that widespread diffusion of GIS to NGOs is likely to increase conflict among groups may be negative, we should recognize that in democratic societies conflict is not necessarily bad. In the first place, open disagreement provides evidence that democracy is truly working, that at least some of the many voices of the people are heard. When many voices are heard, working out a resolution that is acceptable to all is at least possible, although still often difficult to achieve. At the very least, each group will have had an opportunity to air its concerns. When groups do not have a voice, democracy is not working.

The conflict may be lively, but there is no reason to assume that it will necessarily become a disintegrating force, in spite of evidence from the confrontational political programs on cable TV. In fact, some authors (North, Koch, & Zinnes, 1960; Pondy, 1967) suggest that conflict serves an important integrative function, essentially agreeing with Habermas's theory of communicative action. The integration of disparate views comes about as a result of allowing all voices to be heard, thus facilitating a resolution that is at least somewhat acceptable to all parties. In a democracy, public hearings and the judicial system provide arenas in which to raise issues. Day-to-day arenas such as the free press, the Internet, and other mass media provide additional opportunities for interest groups to speak their minds, build consensus, and advocate.

The summer 2003 success of a grassroots movement that used the Internet to motivate individuals to encourage their representatives in Congress to roll back Federal Communications Commission (FCC) rules permitting media giants to grow even larger is an example (Roberts,

2003). Even more encouraging is that this movement included groups that are normally not allies, and, in most instances, are enemies. Among the groups promoting the successful effort were the National Rifle Association, the National Organization for Women, the U.S. Conference of Catholic Bishops, and "dozens of other groups of all stripes and political persuasions" (Roberts, 2003). While there is no reason to believe that these groups have forged a long-term alliance, this short-term alliance provides some evidence to support the notion that conflict (in this case, of many groups and individuals against a common opponent) can serve an integrative function, even if only on a temporary basis.

Because democracies have a variety of arenas in which interest groups can raise their concerns, the likelihood of conflicts becoming massive and negative is relatively small. Still, from time to time, violent protests will erupt. However, historically, these protests are the result of long-brewing inequities and conflicts, separate from the introduction of technology. Overall, the potential benefits to a democracy outweigh any foreseeable disadvantages.

Conclusions

Although earlier hypotheses suggested that GIS will help to ease conflict among groups by providing more and "better" information, the likelihood of this outcome is difficult to support. On the contrary, as GIS proliferate and "trickle down" to more groups, we maintain that conflict is likely to grow. This will occur largely as the result of GIS's inability to influence interest (value) conflict in any significant or direct way. Equally important is GIS's potential to increase cognitive (fact-based) conflict by making more data available to more groups and individuals, who can rearrange them or analyze them in a variety of ways in order to generate new information or knowledge. This new information can, in turn, be used to support arguments in support of alternative policy goals, and thus will tend to increase conflicts.

The case of the proposed Bailly nuclear generating facility is an example of how fact-based arguments (cognitive conflict) may be used to promote a particular interest. As we have seen, the judicial system provides an important forum in which cognitive conflict often serves as a surrogate for interest conflict. Any technology that makes available more information that can be used to support cognitive conflict is a valuable tool for a group that seeks to derail an unwanted land use or promote an alternative land use and has the knowledge to mount a public or legal campaign. If an NGO is wealthy, so much the better. When a delay is the next best thing to a clear-cut victory, organizations will find it to their advantage to bring more facts to the table in order to support their policy objectives. If one set of facts is not sufficiently persuasive, perhaps another will do the trick.

Little has been done to answer our questions about the role of GIS in conflict resolution. It is clear, however, that the proliferation of GIS as a tool in exploring land-use issues promises to raise this issue as a matter of practical interest. So far, the raw case study evidence suggests that many NGOs are harnessing GIS to promote their policy objectives, with a special eye to the importance of data (facts) to achieve their policy goals (values). Further study is needed to determine to what extent the use of GIS as a tool for policy advocacy will strengthen conflict, and to what extent it will help to promote democracy. In the meantime, NGOs may find inspiration from other groups that have already harnessed GIS as a tool for policy advocacy.

Chapter 12

Ensuring the Qualifications of GIS Professionals

The continuing rapid diffusion of GIS to a wider audience of end users has raised concerns within the GIS community about assuring the competency of GIS professionals. The reasons for concern are as varied as the people who express them. Individuals who aspire to careers as GIS professionals have been eager to have a way to demonstrate their competency to prospective employers. Human resources departments of organizations that hire GIS professionals (especially small organizations hiring their first GIS specialists) have wondered what job titles and qualifications to list in descriptions of position openings and how to evaluate the applicants. Some postsecondary educational institutions have pondered how to provide a sound GIS curriculum for their students, while others have developed GIS certificate programs to take advantage of the demand for GIS education. The international development community has also expressed concern about the possibility that public officials in developing countries may fall prey to unethical purveyors of GIS. In short, the development of standards for GIS professionals and educational programs is an idea whose time has come.

Fortunately, since the earlier edition of this book, there has been a great deal of progress in developing objective criteria for evaluating the qualifications of GIS professionals. Many postsecondary educational institutions have established in-house certificate programs, although they vary widely in their approaches and rigor. The University of Minnesota has instituted a master's degree in GIS. Professional organizations, including the GIS Certification Institute (GISCI) and the AGI (Association Geographical International) have established portfolio-based GIS certification programs. Moreover, the University Consortium for Geographic Information Science

(UCGIS) continues work to develop a model curriculum to serve as a guideline for GIS programs at postsecondary educational institutions. Some view this (perhaps erroneously, perhaps hopefully) as a first step toward an accreditation program. In short, there is a strong push for nationally accepted standards for GIS professionals and educational programs.

This chapter discusses certification and accreditation as means to help ensure the qualification of GIS professionals, and details the current status of both these initiatives. Since these initiatives are relatively new, their long-term results are unknown at this time. Still, they are useful in providing guidelines to organizations that must hire GIS practitioners, as well as to individuals who are eager to demonstrate their expertise in GIS.

Demand for GIS has grown rapidly in the 1990s and into the 2000s, and the technology is proliferating among public, private, and nongovernmental organizations. One of the consequences of this increased demand is a difficulty finding GIS professionals who are competent both in the appropriate use of the technology and in the content area of the employing organization. Some fear that unethical individuals will jump in to take advantage of employment and business opportunities. The employment of inadequately trained or unethical individuals in the GIS field has the potential to cause the failure of GIS projects, which in turn has the potential to damage the credibility of both GIS and the profession itself.

As a result, there has been a growing concern within the GIS community to find a mechanism to assure the competency and ethical behavior of individuals who present themselves as GIS professionals and to help new GIS employers identify appropriately qualified individuals. This concern resulted in a great deal of activity in recent years to develop objective standards for aspiring GIS professionals as well as for educational programs in GIS. These efforts are now bearing fruit.

Before continuing, it is necessary to note the difference between two similar means to ensure the qualifications of professionals: certification and licensure. Both strategies rely on first establishing a core body of knowledge for the profession or field that is seeking to provide a mechanism for evaluating the competency of individual practitioners in the field. In many cases, both certification and licensure rely on either testing or evaluation of a portfolio to assess the qualifications of individuals. What differentiates licensure from certification is that licensure takes the added step of gaining the support of state (and sometimes federal) legislatures that pass laws to require that individuals pass specific professional tests or portfolio reviews as a prerequisite for practicing in the field. Certificates are administered by professional organizations on their own terms, while licensure is a matter of state (or sometimes federal) law. Professional organizations are often instrumental in developing and implementing tests required for state or federal licensure. However, licensure or certification is not the only mechanism designed to assure the competency of professionals. Accreditation of programs of instruction in a field is also a well-accepted means to this end.

The most commonly suggested strategy to ensure the qualifications of GIS professionals has been the development of a certification program for individuals who wish to hang out their GIS shingle. In fact, the National Council of Examiners for Engineering and Surveying (NCEES), which employs government-sanctioned licensure to assure competency among its members, made inroads in Georgia and California toward the achievement of this goal beginning in the early 1990s. By the year 2000, the NCEES had drafted a model law that it hoped to implement throughout the United States via passage by individual state legislatures. This model law would have limited the handling of any GIS data measurement or editing that would have had an impact on official base maps to licensed NCEES land surveyors.

As a practical matter, the question of competency warrants serious discussion. Thus, it is an important matter for organizations implementing GIS. In some instances, organizations will hire GIS professionals from outside the company or agency. In other instances, the organization will train existing in-house personnel. Either way, competency and ethical behavior are serious matters. In the first case, human resources (HR) departments must decide among a pool of applicants who will best fit the organization's needs. Guidelines drawn from certification requirements may be valuable in helping to assess and rank applicants. In the second case, HR departments must decide which skills their employees must learn in order to gain competence. Guidelines drawn from a GIS model curriculum will help organizations to choose appropriate education and training options.

This chapter discusses both certification and accreditation as alternatives to a professional norm of peer pressure, which, by default, was the model the GIS community had previously relied on to ensure the qualifications of its members. The chapter begins with a brief discussion of the current situation in the GIS community, then discusses two alternatives, certification of individuals and accreditation of programs, as means to assure the competency of GIS professionals. The discussion of certification includes experiences from two other professions: medicine and planning. This chapter continues by recounting recent developments in GIS certification and accreditation. An important part of the development of certification has been the development of a code of ethics for GIS professionals, which is also covered. The chapter concludes with a discussion of where these developments might lead.

Background

The goal of assuring the competency of GIS professionals is a noble and laudable objective. It is also necessary as a means to assure and protect the credibility of both GIS as a technology and individuals who call themselves GIS professionals. This issue is therefore significant to the GIS community.

It would be false to assume that because the GIS community has only recently put in place a professional certification process that there has been no mechanism in place to assure the competency of professionals in the field. The original mechanism to ensure the qualifications of GIS professionals was basically an honor system that relied on both formal and informal networks of individuals and organizations to promote an atmosphere of integrity and ethical behavior. A number of well-respected individuals who were instrumental in the creation and development of GIS set an ethical tone for the community as a whole. In addition, several organizations (the AAG, ACSM, AM/FM International, ASPRS, and URISA) have developed a close working relationship with the GIS community, providing an additional source of norms and conventions of professional behavior.

In the early years, this informal default arrangement worked reasonably well to promote both competent and ethical professional behavior. A reason why this model worked is because through the 1980s, and into the 1990s, the GIS community was relatively small in number. Early on, peer pressure and word-of-mouth assessments were adequate to promote competency and ethical behavior. Keep in mind that through the 1980s a large number of individuals joining the ranks of GIS professionals were able to meet many of the profession's founders, and in many cases to establish an ongoing professional relationship with one or more of them. In fact, like many others, both authors of this book have been beneficiaries of substantial professional encouragement and support from many of the profession's founders. Ethicists would describe this situation as providing examples in "virtue ethics."

The ability of so many newcomers to develop ongoing professional relationships with these founders allowed the newcomers to observe the high standards of competency and ethical professional behavior—not to mention the enthusiasm—of the founders. Newcomers were thus encouraged to hold to these standards, and felt that it was in their best interest to comply with the high standard of competency and ethical behavior that their mentors set. It is also useful to bear in mind that during the first quarter century of GIS (1964–1990), there were relatively few GIS implementations. Most of these were custom-built, many of them either by one of the GIS founders or one of their students.

As GIS became more well accepted, and as more organizations with a wider variety of missions and experiences implemented the technology, the GIS community grew in size. In recent years, fewer newcomers have had direct personal access to the profession's founders and thus have been unable to observe them personally. More newcomers were attracted to the industry because of good salaries, with little regard for the history of GIS, direct contact with its founders, or knowledge of its norms and conventions of professional behavior. Eventually, peer pressure and word-of-mouth reputation became inadequate regulatory mechanisms for the profession. It was therefore both wise and necessary for the GIS community to develop formal mechanisms to maintain the quality and credibility of professionals in the field.

Expertise as a Foundation for Certification and Accreditation

To fully understand the concepts of certification and accreditation, it is useful to explore their roots, which are found in the ideas of disciplinary or professional expertise. Much of what we know about expertise derives from Max Weber's theory of bureaucracy (Weber's books were originally written at the beginning of the 20th century, but first translated into English only at midcentury).

Webster's Dictionary defines *expertise* as "specialized knowledge or skill: mastery." Expertise is important in many aspects of life, from dog training to neurosurgery to GIS. Each field determines for itself which particular mix of specialized knowledge and/or skill set is required to demonstrate competency within that field. The process by which a field determines the knowledge and skills that are crucial to demonstrating competency varies from field to field, but it typically involves an evolution from an informal network of specialists discussing a variety of topics relevant to the field to a formalized organization with codified rules. When the first edition of *Managing Geographic Information Systems* was published in 1994, the GIS community had just begun to make the transition from discussion of what makes GIS a field to its status today, where the body of knowledge has become more formalized.

Expertise forms the backbone of professionalism; this is why it is so important. Professionalism is possible within specific fields largely because of the unique combination of information and knowledge that each field claims as its own. Fields that cannot achieve consensus on exactly what mix of specialized knowledge is required to demonstrate expertise in the field have been known to question their own validity as a profession. Normally, this occurs within professions with highly diverse missions and members. Ten years ago, there was little faith that the GIS community could agree on a central body of knowledge in the field. Today, there is far greater consensus that this is, indeed, possible. In fact, the coalescence of an identifiable, multidisciplinary community focusing on "geographic information science," as described in Chapter 2, has been a key prerequisite for achieving this consensus. The UCGIS is playing a critical role in the ongoing development of a GIS body of knowledge (*www.ucgis.org*).

The size and complexity of a profession's body of knowledge varies dramatically from field to field. GIS encompasses a relatively broad body of knowledge that is also highly complex. The spatial component of GIS, in particular, is highly specialized and outside the range of common knowledge. Furthermore, the integration within GIS of base maps, which include property boundaries and property ownership information (which itself confers other important rights; see Clark, 1981), adds another dimension to the complexity of GIS expertise.

Along with serving as the backbone of professionalism, expertise is often cited as the most crucial prerequisite for the development of certification and accreditation processes. Agreement on the core expertise of a profession must come first, followed by the development of a process to determine if an individual has mastered this core knowledge. These are the steps that a profession must put in place to assure that people who claim to be an expert within a particular field actually possess the necessary expertise to make this claim. Certification of individuals and accreditation of educational programs thus is designed in part to set a standard

of competency and ethical behavior for specific fields. Indeed, this rationale has been a driving force for development of a certification process
for GIS professionals, and also for the crafting of a model curriculum for
GIS educators.

While the development of expertise may serve positive purposes
(e.g., setting a standard of competency), Cayer and Weschler (1988: 45)
also note that the expertise of professions and the concomitant control
over information may lead to a concentration of power within the profession. Similarly, Habermas (1970) suggests that experts may use their specialized knowledge to build a technocracy, thus gaining hegemony within
their field. Weber (1968b) raises similar concerns about the elevation of
technical experts to the status of a mandarin caste, as happened in
ancient China. He notes that many professions gain a virtual monopoly in
their area of expertise, which makes it very difficult for outsiders to evaluate the performance of members of the profession. Likewise, Cayer and
Weschler (1988: 45) note that the expertise of professions and their concomitant control over information by professionals in their field often
leads to a concentration of power within the profession. The American
Medical Society (AMA) is a prime example of a profession that has used
its expertise to consolidate control over the practice of medicine in the
United States. In fact, Berlant (1975) goes further, suggesting that the
AMA's control over medical licensure has enabled this group to transform
itself into a medical monopoly in the United States.

In extreme cases, the concentration of expertise that is necessary to
assure competency within a field and the use of expertise can lead to creation of a technocracy. Given the technical nature of GIS and spatial analysis and modeling, as members of the GIS community, there may be reason for some concern about the creation of a GIS technocracy.

For example, in the United States, professional land surveyors, represented by the American Congress on Surveying and Mapping (ACSM)
and by state-level organizations of "professional land surveyors," have
attempted several times to make surveying the dominant training for GIS
professionals in the United States. Initially, in the 1990s, land surveyors
lobbied state legislatures throughout the country to try to pass legislation
that would require that *only* a licensed land surveyor would be permitted
to modify base maps used in official GIS. Had the group been successful,
this initiative would have established land surveyors as a GIS technocracy,
while potentially leaving GIS professionals trained through geography,
planning, and other related fields out in the cold. As we go to press in
2007, the Management Association for Private Photogrammatic Surveyors

is once again lobbying to make land surveying the required credential for GIS professionals.

Certification and Licensure in Two Other Fields

Many professions, including medicine, law, surveying, planning, and barbering, have long-established certification or licensure procedures. These procedures are characterized by varying degrees of effectiveness. This section describes the successes and failures of procedures in two of these fields as a means to provide insight into what certification might mean in GIS.

The concepts of professionalism and certification imply internal, self-regulatory and self-promotional processes. It is within this context of both self-regulation and self-promotion and the reinforcing functions of these elements that we understand the effectiveness of certification or licensure. Effectiveness of professional certification may be defined in two ways: first, as the competent, appropriate, and ethical behavior of the members of a specific profession in meeting their responsibilities to the larger society (self-regulation); second, "effectiveness" may refer to the success of members of the profession in the marketplace, and the ability of the profession to ensure that practitioners it has certified are hired in preference to practitioners who do not have the profession's official seal of approval (self-promotion). An examination of certification in the medical and planning professions helps to illustrate these points.

Development of a Professional Monopoly: The American Medical Association

One of the most effective of all professional organizations in the United States is the American Medical Association (AMA). Formed in 1847, its raison d'être was to upgrade and maintain educational and professional standards in medicine (Berlant, 1975: 226). At the urging of Nathan Smith Davis, whose ideas formed the basis for early AMA policy, the AMA adopted a system that separated teaching and licensing. Medical schools continued to function as they always had, while licensing was instituted at the state level, with medical societies having responsibility for appointing individuals to the state licensing boards. In addition, the AMA established a policy that required *both* a diploma *and* a license in order to enter the practice of medicine. Previously *either* a diploma from a medical school *or* a license had been sufficient to allow a doctor to set up a practice (Berlant, 1975).

"Protection of the public against quacks," Berlant (1975: 227) notes, was the primary justification for the development of what eventually became a monopoly in medical services by the AMA. Following in the German tradition of state licensing superimposed on university examinations, Davis effectively promoted legislation that ultimately pressured medical schools into a specific line of development, what Berlant describes as "orthodox medicine" (227). Orthodox medicine in this case refers to the medical theory and practices of the AMA. The imposition of this legislation effectively made AMA's brand of orthodox medicine the *only* medicine that would receive the official seal of approval of state legislatures across the United States. Thus, AMA medicine effectively became the sole basis for accreditation of medical programs as well as the sole basis for licensure examinations for aspiring physicians throughout the United States. What had previously been professional dominance by the AMA thus became hegemony. With the support of state legislatures and the placement of AMA-approved physicians, who began to fill the ranks of state medical boards, the AMA's hegemony became a virtual monopoly over the practice of medicine in the United States (Berlant, 1975).

In the years after its establishment, the AMA became synonymous with quality health care in the United States. Practitioners and products alike proudly displayed the AMA's seal of approval. In recent years, however, concern about monopolistic domination by the AMA has arisen. Complaints about the arrogance of doctors have become commonplace. Until very recently, one dared not question the medical judgment of the physician: the doctor alone possessed the medical expertise to save a life. The AMA had come to resemble the technocracy described by Habermas (1970). Berlant (1975) suggests that the legislative seal of approval that licensure gave played an important role in the evolution of the AMA from a professional authority into an exceedingly powerful monopoly.

We might logically conclude that the complexity of medicine itself, along with the universal importance of medical care, were critical factors in the evolution of the AMA's dominance. Certainly, few state legislators possessed the medical education and experience to regulate the medical profession. They therefore willingly accepted the expertise of the AMA's representatives to help them with this critical task by allowing the AMA to choose members of the state medical licensing boards. The help from AMA physicians in regulating health care in the United States began with setting up the system, and the help continues with ongoing participation of AMA physicians on state medical boards.

Voluntary Certification:
The American Institute of Certified Planners Experience

If the AMA stands out as a professional organization whose licensure process assured a virtual monopoly for its members, then the American Institute of Certified Planners (AICP) is an example of what happens when certification is a totally voluntary exercise.

In his discussion of the AICP certification examination, Rasmussen (1986: 7) argues that "certification by the professional society means planners themselves control their own occupational standards," thus implying that the planning profession's voluntary system promotes maximum autonomy for the AICP. However, comparing the AICP to the AMA suggests that there is a price to be paid for this autonomy.

In contrast to the AMA, the planning profession is open to any individual who cares to call him- or herself a planner, regardless of certification. The good news is that there is a statistically significant difference in the salaries of professional planners who are certified by the AICP compared to those who are not (Morris, 1996).

While certain jobs in planning (most commonly academic and management positions) require that candidates have AICP certification, most positions do not. Many positions advertised in the American Planning Association's *JobMart* do not even require completion of a degree program in planning.

If we consider these two facts together, we might draw a conclusion about the value of AICP certification that is at odds with Rasmussen's (1986) assessment. The openings for planners that required certification tended to be either in academia or in higher ranking positions. Therefore, while AICP certification may not be necessary for entry-level employment in the planning profession, certification may lead to promotional opportunities. It is possible that the higher salaries that Morris (1996) discovered in her research were the result of the promotion potential of AICP-certified planners.

From the perspective of the planning professional, however, the value of AICP certification seems limited, at least as far as entry-level employment prospects are concerned. Since few planning positions are set aside exclusively for AICP planners, individuals considering making the effort to earn certification must weigh the potential value of certification as a job applicant against the time, effort, and expense required to achieve certification. If all (or even most) planning jobs required AICP certification, the decision would be more clearly in favor of pursuing certification. On the other hand, the fact that there is a correlation between AICP certification and higher salaries certainly should get the attention of planners who wish to remain in the field and who aspire to management positions in planning. In addition, the decision to earn AICP certification is often one based on personal and professional pride.

Because a significant proportion of planning jobs are not held exclusively for AICP planners, one could logically challenge Rasmussen's (1986) assertion that the AICP exercises autonomous control over the quality of planning practitioners. In fact, if one need not achieve AICP certification in order to work as a planner, and if indeed *most* planning positions do not require AICP certification, then it is clear that the AICP exercises little quality control in the planning marketplace. Rasmussen himself admits that only about 20% of the members of the American Planning Association (APA) are also certified members of the AICP (and, by logical conclusion, have passed the AICP certification examination). Presumably most members of the APA are practicing planners. Based on Rasmussen's 20% certification figure (or 80% uncertified figure), we can logically conclude that relatively few people holding down professional planning positions have documented their competency in planning. This is not to say that most planners are unqualified, but rather that neither they nor the AICP has documented their competency.

To reiterate, the AICP certification procedure is intended as a means to assure quality among planning practitioners. It is probable that planners with AICP certification possess high-quality skills and experience. However, in practice, it would appear that the AICP exercises little control over the practice of planning in the United States. That is to say, the AICP has relatively little control over who works as planners in the United States. Without some mechanism (either voluntary or coercive) to limit entry into planning practice exclusively to AICP members, the organization is unable to assure quality across the board within the planning profession. Governmental regulation (licensure) is not necessarily the only means to ensure that only professionals who have demonstrated their competence can hold jobs in the field. If a professional organization is successful in persuading prospective employers to hire only certified professionals, this is another way to achieve the same goal. Apparently, the AICP has not yet been able to persuade most employers of planners that AICP certification is a crucial indicator of competency.

The differences between the AMA, which has a virtual monopoly on medical practice in the United States, and the AICP are readily apparent. In the first place, the approaches of the two groups are very different, as the AMA assiduously courted legislators in its early stages to assure the dominance of its practice of medicine over all other types through licensure. The AICP, by contrast, has resisted what it perceives as the threat of legislative control of its profession. In fact, it appears that the AMA controls the legislature within its area of expertise, not the other way around. The outcomes are dramatically different: the AMA has achieved a stranglehold on medical care in the United States; the AICP credential is a lofty goal to which many planners may aspire, but which a relative few actually attain.

Accreditation: A Brief Description

Accreditation differs from certification in that its focus is at the institutional level rather than the individual level. Accreditation has a long history in academia, with colleges and universities earning accreditation as a whole, and with specific academic programs earning accreditation within their specific professional or academic sphere. In the case of GIS, the latter approach is relevant.

Generally, the first step in developing an accreditation process is the establishment of a specific set of evaluation criteria. These criteria generally include standards for personnel (both quantity and quality), facilities and equipment, and coursework. They are normally set by an organization that has credibility within the profession. For example, the AICP sets the standards for accredited planning programs at the university level. The National Association of Schools of Public Affairs and Administration sets the standards for accredited programs in public affairs and administration. The AMA sets standards for medical education in the United States.

Once evaluation criteria and standards are in place, programs apply for accreditation, initially providing written documentation as required by the accrediting organization. If the initial application and materials are approved by the professional association, the accrediting organization sends a site inspection team to examine the premises of the program requesting accreditation. Providing that the inspection team is satisfied that the program (including faculties, libraries, facilities, equipment, etc.) meets or surpasses the evaluation standards, the program earns accreditation for a predetermined period of time. When that period of time expires, the program must begin the accreditation process anew. However, if in the meantime conditions within the program change in areas covered by the evaluation standards, the program risks losing its accreditation. This has been occurring in planning programs, which are required to have at least five full-time equivalent (FTE) faculty members to achieve and maintain AICP accreditation. In the current climate of university downsizing, faculty loss is becoming more common, jeopardizing some planning programs on the basis of too few faculty members.

Like certification, implementation of accreditation procedures runs the risk of encouraging a technocracy, and therefore must be carefully considered. In general, accreditation may have certain advantages over certification. First, because there are fewer colleges, universities, and private providers offering GIS courses and programs of study than there are individuals who would aspire to become GIS professionals, there would be fewer schools seeking accreditation than individuals seeking certification. Managing the process could be somewhat easier because of the smaller numbers. In a similar vein, the familiarity of members of the GIS community with one another and with one another's programs may facilitate accreditation reviews.

The GISCI GIS Certification Program

Background

January 2004 brought with it the formal official rollout of a GIS Certification program sponsored and administered by the GIS certification Institute (GISCI, at *www.gisci.org*). This section discusses the process by which this program came into being and the requirements for certification.

At the time when the first edition of *Managing Geographic Information Systems* was published, certification of GIS professionals had fallen off the radar screen. The earlier incentive for discussion, efforts by the surveying profession to require licensure for GIS professionals, had been largely unsuccessful by the mid-1990s. At the end of the 1990s, however, the group tried again to promote a model law that would have required that only land surveyors licensed by the NCEES be permitted to use GIS to create or alter official base maps.

Concerned about keeping open employment opportunities for GIS specialists generally, other members of the GIS community began working together to address this development. One avenue for addressing the situation was to learn more about the NCEES model law and to discuss their concerns. Another avenue was to form a committee to discuss a nationwide GIS certification program in the United States. The Urban and Regional Information Systems Association (URISA) took the lead in both initiatives. The GIS Certification Committee that URISA established was deliberately structured as an open group with members from any and all relevant organizations with an interest in GIS. Membership in URISA was *not* a requirement for membership on the GIS Certification Committee and participation in the committee's work.

Learning more about the NCEES model law enabled nonsurveyor members of the GIS community, represented by the URISA-based committee, to discover that the model law was somewhat more limited in scope than they first feared. In addition, the GIS Certification Committee established ongoing communications with the NCEES in order to keep abreast of any developments. At the same time, it became clear that there was a growing need for guidelines for GIS professionals, and more members of the URISA Certification Committee expressed a willingness to support establishing a certification program. During 2002 and into 2003, members of the committee worked to develop a portfolio-based certification procedure based on the process established by the United Kingdom's Association of Geographic Information (AGI). Beginning in the spring of 2003, the GIS Certification Committee implemented a test run of its certification program for a group of applicants from the State of Georgia. Twenty-nine of these applicants achieved certification. The GISCI Certification Program officially opened for regular business on January 1, 2004.

The Elements of GISCI Certification

The certification program established and now implemented by the GIS Certification Institute holds the hope of providing help to organizations seeking to hire GIS professionals. This help comes in two forms. First, organizations will be able to hire GIS professionals who have earned certification. Second, organizations may also use the GISCI requirements for certification to help them develop their own GIS job application review criteria.

There are three major components to the GISCI certification program: (1) educational achievement, (2) professional experience, and (3) contributions to the profession (GISCI, 2007). Applicants for the program must accumulate a total of 150 points, with a minimum number of points in each of the three categories in order to earn certification: 30 points in education, 60 points in experience, and 8 points in contributions to the profession. If you have been doing the math, you immediately notice that applicants must earn at least 52 additional points in order to reach the 150 point total needed to become certified by GISCI. As the official GISCI website points out, "The additional 52 points can be counted from any of the three categories. The applicant has complete flexibility in deciding how to make up this difference. In other words, the 52 points can be made up from any combination of points from any one (or more) of the categories" (GISCI, 2007).

Applicants may earn points in the education category through three avenues. The first is "credential points." Applicants earn credential points through successful completion of a formal degree or certificate program offered by an accredited institution. The highest number of points available under this education category is 25 (for a master's degree or PhD in a GIS-related field). Bachelor's degrees and completion of a GIS certificate, such as those offered by a growing number of accredited institutions of higher learning, are also eligible for credential points.

The second way to earn points to fulfill the education component of GIS certification is through "course points." Applicants earn these points by successfully completing individual courses, workshops, and other formal, documented educational activities in GIS. Courses that qualify as GIS follow guidelines established by UCGIS as part of its model curriculum. These are described in detail below, in the section on the UCGIS Model Curriculum.

The final way that applicants may earn points toward the education component is through conference attendance. Applicants earn these points "in recognition of the valuable informal learning afforded by participation in [GIS-related] meetings and conferences sponsored by professional societies and regional and local user groups" (GISCI, 2007). The "informal" nature of the learning provided by conference attendance is reflected in the relatively small number of points that an applicant may earn for each conference. Normally, applicants for GISCI certification will accumulate points in at least two of these three categories. Most applicants earn points in all three.

The second category in which an applicant must earn points to be eligible for GISCI certification is in the area of professional experience. The GIS Certification Institute identifies job experience as

> the most important factor contributing to an individual's qualifications as it
> allows opportunities to become skilled at the application of GIS technology

to real world problems. Failures and successes in these contexts provide valuable learning experiences that, in turn, allow growth and expansion of skill sets. In addition, the professional working environment, where one is often working with other GIS professionals who have different skill sets and different experiences, provides opportunities to gain knowledge from one's peers. Successes, failures, and access to mentors all form skill development opportunities in the working experience, and the longer one is exposed to these opportunities, the more one is qualified to address new problems. **There-fore, four years of experience is the minimum number of years required for GIS Certification."** (GISCI, 2007; bold and underlining in the original text)

The GISCI also makes a distinction between various types of work in GIS professional positions. Individuals who have held positions in data analysis, system design, programming, or a similar GIS position receive the highest number of points per year on the job (25). GIS professionals who have held positions in data compilation, GIS teaching, or a similar position receive fewer points for each year on the job (15) than the first group, but more points per year on the job than GIS professionals whose duties have been limited to using GIS (10). GISCI also gives 10 bonus points to GIS professionals who have served in supervisory or management positions within the context of their GIS employment.

The minimum number of experience points required for certification is 60. The minimum number of years of professional service that one must have in order to become certified as a GIS professional is four.

This brings us to the issue of "grandfathering." The GIS Certification Institute recognizes that many people working in GIS today did not have the benefit of GIS coursework, but have worked in the field for a number of years. In order to accommodate these individuals, GISCI developed its "Grandfathering Provision." Specifically, the Grandfathering Provision allows the experience of highly experienced applicants to compensate for deficiencies they may have in education and/or contributions to the GIS profession (discussed a little later). The GISCI Grandfathering Provision deems applicants eligible for this special dispensation if they have held a GIS position in data analysis, system design, programming, or a similar post for a minimum of 8 years. GIS professionals who have held positions in data compilation, teaching, or a similar position must have held that position for a minimum of $13\frac{1}{3}$ years in order to be eligible for grandfathering. If a GIS professional has served in a user position, he or she must have been employed in that post for at least 20 years before grandfathering is a possibility. Applicants may earn GIS certification using the Grandfathering Provision only for the first 5 years of the program (January 1, 2004–December 31, 2008).

The final category in which an applicant for certification through the GIS Certification Institute must earn points is called "contribution to the profession." The rationale behind this category is that "it must be recognized that professional contributions in the form of conference planning, publications, committee/board participation, outreach and other related efforts are fundamental to the health of any profession" (GISCI, 2007). GISCI notes that if there is a professional norm that values and rewards contributing to the profession, then employers of GIS professionals may be encouraged to support the participation of their employees in these important activities.

Six separate categories are included in the GISCI contribution point schedule. The first is GIS publications, defined as writing or reviewing certain GIS-related material. Membership in professional associations with a GIS focus is also eligible for "contribution to the profession" points. The third category is participation in a GIS conference, which would include organizing a portion of a conference or making a presentation at a local, state, or national GIS conference. The fourth category is GIS workshop instruction, which includes presenting a workshop at a local, state, or national event (provided that the presentation is not on behalf of the applicant's employer). Applicants may also receive points based on GIS awards that they have earned. Finally, other GIS contributions, including organizing or participating in GIS day-type events, are also eligible for points under the "contributions to the profession" point schedule. In order to achieve certification, a GIS professional must accumulate at least 8 points for their contributions to the GIS profession.

An important element of GISCI certification is the GIS Code of Ethics. In 1992, Will Craig was the first to raise the issue of a code of ethics for GIS professionals, and he was instrumental in crafting the text for the GISCI Code of Ethics. Each and every GIS professional who earns certification through GISCI must sign the GIS Code of Ethics.

UCGIS Model Curriculum/Body of Knowledge

UCGIS has become an important multidisciplinary group that has taken the initiative to develop a model curriculum for GIS programs at the postsecondary level. While the emphasis in this model is to provide guidelines for the GIS academic community, potential employers may also use these guidelines as they review credentials of prospective GIS employees.

The UCGIS Body of Knowledge provides a detailed discussion of the knowledge that a GIS professional must possess in order to carry out his or her responsibilities. It is worth visiting the UCGIS website (*www. ucgis.org*) in order to review these recommendations in detail to see how they apply to any particular application of GIS.

What to Do in the Meantime

We conclude this chapter by reminding the reader of the original point: there is a growing need to assure the competency of GIS professionals. This point will certainly not be lost on anyone who has had to hire someone to help implement their GIS. There are several things that potential employers can do to assure the qualifications of the GIS professionals they hire.

First, read as much as possible about GIS and your specific applications. This will help you to understand your needs more clearly and to identify the skills and capabilities you require. Moreover, it will help you to identify organizations that have accomplished similar implementations with which you might network. Reading will also help you to learn more information about the education in GIS that is available at various institutions of higher education.

Second, network. Before you hire someone, attend local, regional, and—if possible—national GIS conferences and workshops. This will give you an opportunity to meet potential employees, either directly or through contacts that you might make. In many cases, formalized procedures enable prospective employers and prospective employees to meet for face-to-face interviews during the course of the conference.

As part of your networking activities, talk with other participants, the exhibitors, and presenters about your needs. Solicit their advice. These conversations may either identify specific people who may be right for your opening, or at the very least help you to identify the characteristics you should be looking for. This will help you to develop a position description. These networking activities should also help you identify specific avenues by which you can advertise your opening.

Third, remember that vendors are a valuable resource. Particularly if you have already chosen your GIS, you will find your vendor of help at this phase. Vendors have a powerful economic interest in the successful implementation of their GIS in your organization. The vendor most likely has formalized training programs (in-house or sometimes on-site) that your potential employees should have completed. In these situations, the person who has successfully completed such a training program will often have earned a certificate of completion.

In addition, vendors also tend to have extensive contacts in business, in the public sector, and in academia. They may be aware of programs that produce GIS professionals that fit your specific needs. They may have heard of someone who is looking for a new professional challenge. Vendors are an important link in any network.

Conclusions

The rapid pace at which organizations are adopting GIS means that we must not become complacent. The demand for more and more professionals means that identifying and hiring qualified professionals may become more difficult in the short run. In the long run, we can expect market forces to begin to blunt any shortages. Increasingly, institutions of higher learning are implementing GIS courses and programs that will produce more professionals; many of these follow the recommendations of UCGIS.

Organizations seeking to employ GIS professionals should take the usual precautions that they would take in hiring any employee. Verify resumes, check references, and hold careful interviews (multiple interviews, if need be). These practices, along with those described earlier, will make the difference. In addition, however, organizations will also benefit from gaining a thorough knowledge of GIS and the needs they serve.

Chapter 13

Legal Issues in GIS

When the first edition of *Managing Geographic Information Systems* came out, a number of respected legal scholars were already examining legal issues related to GIS. At that time, however, because the mass diffusion of GIS technology was in its early stages, their discussions of legal issues tended to be topic by topic. Today, that has changed. In addition to the topical discussions of GIS and the law, more comprehensive and integrated studies and expositions of legal issues in GIS exist. For example, Cho's (1998) book describes developments in the English-speaking world (especially Australia and the United States) with respect to GIS law, while Onsrud's "GIS and the Law" (2005) course is available online, as well as in-person at the University of Maine. Because the discussion of GIS law has matured into a more comprehensive body of knowledge, it is both necessary and desirable to include a chapter on legal issues concerning GIS in this second edition.

Using the comprehensive treatments of GIS law by Cho (1998) and Onsrud (2005) as the basis of our framework, we begin this chapter by laying out the issues of greatest concern in GIS law. We proceed to discuss each issue, first defining it, then discussing the mix of statutes, case law, and other relevant trends. We conclude by discussing prospects for the future of GIS law.

An Overview of GIS Law

The body of knowledge associated with GIS and the law has advanced dramatically in the past 10 years, paralleling the development and diffusion of GIS. Twelve years ago, Obermeyer (1995a) prepared an overview of legal issues in GIS that relied heavily on short articles in GIS trade publications because these were virtually the only sources available at that time. The topics discussed in those articles could be placed in three categories: (1) property rights regarding geographic data and information (including sales and other remuneration for providing data), (2) liability for GIS products, and (3) data privacy rights. Relatively little case law existed at that time, and there were no integrated books providing an overview of GIS law.

Things have changed. The extensive diffusion of GIS to a mass audience has paved the way for an expansion of statutes and case law in the area of GIS, which in turn has translated into the evolution of literature on the subject. On the one hand, the literature has increased in quantity to accommodate the expansion of GIS law in the real world. On the other hand, legal scholars in GIS have accompanied their expansion of literature with efforts to integrate the literature into a more unified body of knowledge. We see this coming together most readily in the evolution of a consensus on key issues in GIS law. The three major legal issues in GIS that Obermeyer identified earlier have become six issues today.

Cho (1998) and Onsrud (2005) have nearly 100% overlap in the issues they identify as crucial to GIS law. The issues on which Cho and Onsrud express consensus are (1) liability (contract and tort); (2) public access, use, and ownership of data; (3) intellectual property; (4) copyright; (5) privacy; and (6) evidentiary admissibility of GIS products. As we will demonstrate, many of these topics are interrelated and sometimes mutually reinforcing or contingent.

This chapter presents an exposition and discussion of each of these issues in the order they are listed above. We begin with liability.

Liability

Laws regarding liability are exceedingly important for organizations that use, sell, or give away geographic information, as well as for purveyors of GIS software. "As a general proposition," says Onsrud, "legal liability is a harm-based concept" (1999: 643). Providers of GIS software, data, and other products must perform their responsibilities competently. The courts have held that if their actions result in mistakes that damage or harm others, such providers should bear some responsibility for these damages.

Appropriate communications, contracts, and good business practices help to reduce liability exposure, but the courts remain the avenue of last resort when all else fails (Onsrud, 1999). Both Cho (1998) and Onsrud (1999) discuss liability within the context of tort law. *Tort* is a French word that means "wrong" (Cho, 1999: 95). A tort is a "private or civil wrong or injury to a person or property that does not arise as the result of a breach of contract" (Onsrud, 1999: 647). Tort law supports two important policies: punishing wrongdoers for their conduct and compensating parties for their injuries.

Onsrud (1999) identifies three traditional classes of tort liability: negligence, fraud, and products liability. *Negligence* is defined as "conduct that breaches a duty of care for the protection of others against unreasonable risk of harm" (Restatement [Second] of Torts, 1964: 282, in Onsrud, 1999: 647). Any person or organization that provides a good or service has a "duty of care" to the parties who receive the good or service; the provider also must meet a "standard of care" when it provides the good or service (Cho, 1999: 97). Embedded in these two concepts is the idea that the care must be consistent with the care that a "reasonable person" would take. Moreover, in order for negligence to exist, the damage caused by such negligence must have been foreseeable (Cho, 1999).

The second category of tort liability is *fraud*. Negligent and fraudulent misrepresentation is one variation of fraud. Fraudulent misrepresentation occurs when a professional with a financial interest misrepresents a fact, opinion, intention, or law in order to persuade another person to act (or to refrain from acting), knowing that the second person relies on the advice of the professional. Within the context of a contract, fraudulent misrepresentation is "only subtly different from the modern law of breach of contract" and thus is very difficult to prove (Perritt, 1996: 239, as reported in Onsrud, 1999).

The second variation is fraud in the inducement of contract. This type of fraud occurs when one party to a contract knowingly provides erroneous information designed to persuade the other party to agree to the contract. Such erroneous information includes fabricated testimonials or test results (Perritt, 1996: 243, as reported in Onsrud, 1999).

The third class of tort liability is strict *products liability*. The idea behind products liability is to keep defective and unsafe products off the market. Products liability laws reduce the burden of proof on harmed parties, thus making the manufacturer and others in the supply chain responsible for damages caused by their unsafe or defective products. Products liability laws keep manufacturers from claiming that they met industry standards for quality control and design under negligence theories, thus providing added protection for consumers (Onsrud, 1999: 249).

At the heart of the discussion of legal liability in GIS is the question of whether geographic information is considered a product, a publication, or a service. Two obvious areas for product liability with respect to geographic information are navigation and aeronautics. In the United States, aeronautical and navigational charts (two commonly used forms of geographic information) have been held to be "products" and thus are subject to product liability laws.

The Information Chain and Liability

Assigning liability in the case of geographic information is difficult because hardware, software, data, and people are all part of a complicated equation. Onsrud (1999) uses the appropriately descriptive phrase "information chain" to refer to geographic information and its handlers that come together in a GIS and its applications. Furthermore, he says, "Use of GIS and data and software inevitably results in some action or decision. If errors or shortcomings have resulted in inappropriate actions or decisions and parties are harmed, the specter of liability arises for dataset and software producers as well as for all other parties involved in handling geographic information" (645).

There are several links in the geographic information chain, beginning with the author, creator, or collector of the data, who originates the data. The next link in the chain is the database producer/publisher, who uses the original data to derive new data or to restructure the database itself. The third link in the chain is the database distributor, who is responsible for the storage of and access to the data. The fourth link in the chain is the search intermediary, who structures the query and retrieves the information. The fifth link in the chain is the end user, who adapts the information and applies it to a specific task. The sixth and last link in the chain is the person responsible for data communications, the person who connects the first five links (Tarter, 1992).

In most cases, the allocation of risk among the six links in the information chain is achieved through formal agreements or other contractual arrangements (Onsrud, 1999). Contractual agreements, however, may be lacking when data are taken from open sources and other free data outlets, which are becoming increasingly common. Certainly caveat emptor applies in all such situations.

Product Liability in GIS

In making a determination on whether product liability laws apply in a specific GIS application, several conditions must be met. First, the manufacturer must be in the business of selling the product in question: economic gain must be a plausible motive. Second, an injured party must be unaware of any defect in the product prior to using it. Third, the manufacturer must know or reasonably expect that consumers would use the product without checking it first for defects. Fourth, the product must be useful and desirable. Fifth, the manufacturer must have taken every reasonable and cost-effective action to make the product defect-free (Onsrud, 1999).

The two most likely scenarios where liability comes into play with GIS are in cases where inaccurate geographic information results in injury or in cases where the misuse of geographic information leads to unforeseen negative consequences. In the United States, legal precedent thus far suggests that where inaccurate geographic information results in injury, the individual or organization responsible for assembling the inaccurate information can and will be held liable (Onsrud, 1999).

The most clear-cut example of a need for assignment of liability in the information chain occurs when geographic information is inaccurate. For example, if a navigation chart gives incorrect water depths, causing a ship to run aground, we can consider the map to be a defective product. In this case, the creator of the map, and any other link that had responsibility for taking the original measurements or for verifying the ground truth of the map, can be held liable.

A more complicated situation arises when a map is misused or used for a purpose other than that which the creator had in mind. For example, if an individual uses a general map for navigation, as opposed to a navigation chart that shows water depths, to help guide his or her ship, this is an example of a misuse of geographic information. In this case, the key question is whether or not the creator of the map has clearly identified the purpose—and limitations—of the map or whether the user should reasonably have been expected to know the limitations on use.

Over the past decade, there have been significant efforts within the GIS community to promote the use of metadata in order to provide sufficient information about spatial data sets so that the inadvertent misuse of data will be less likely to occur. (See Chapter 10 for a detailed discussion of this topic.) Assuming that the metadata accompanying GIS data are adequate, and that potential users of data are aware of the existence of metadata and know how to assess the appropriateness of a data set using the accompanying metadata, they should be able to decide whether or not a specific data set is appropriate. Metadata, however, are not yet universally available and it is not clear that all potential GIS data users know about metadata.

Evolving case law strongly suggests that GIS will be treated like aeronautical and navigational maps, and therefore are subject to product liability laws. For example, the City of Seattle, Washington, was held liable when its map of utilities incorrectly marked the location of electrical utilities. As a result of this locational error, a construction company drove steel piles through electrical lines, causing a major fire in downtown Seattle. The Washington State Supreme Court held the City of Seattle liable as the creator and owner of the map that provided incorrect information. In this particular case, no GIS was in use. However, it is common to use GIS technology to provide maps of public utilities (hard copy or digital) for the purpose of construction activities (Cho, 1998).

One final point regarding liability: Onsrud, Johnson, and Winnecki (1996), citing Johnson and Dansby (1995) and Perritt (1995), suggests that state and local agencies that sell geographic information at a profit most likely face greater exposure to liability than agencies that disseminate the information at cost, according to Freedom of Information Act (FOIA) rules (see discussion below), or provide the data for free.

Public Access, Use, and Ownership of Data

Property rights are at the heart of issues related to law as it relates to the ownership of geographic information. Ownership of property conveys certain specific rights. In the case of geographic information, the owner of the data set also owns the copyright to the information; this ownership confers exclusivity of use and control of the information. In most countries (the United States is the major exception) national governments are allowed to own geographic information and thus are entitled to sell it for profit. In Great Britain, the government sells government data at high prices and achieves nearly full cost recovery. The federal government of France sets its prices relatively low in order to promote the use of its data. In the United States, the federal government cannot legally copyright the data it collects; furthermore, FOIA requires that data sell for no more than the cost to reproduce it, keeping data costs in the United States very low (Obermeyer, 1995a).

In contrast, U.S. local and state governments can legally own data. Some have used the sale of geographic information as a means to generate revenues. In spite of this authority to own data and the implicit authorization to sell it, many local and state governments provide information free or at the cost of reproduction, just as the U.S. federal government does. Governments may use three separate strategies to collect revenues from the use of geographic information by others: sale, lease, and royalty.

Sale, Lease, and Royalty

There are three primary ways by which governments exchange data for money. The first way is through outright sale. This strategy works well for value-added users like a hotel chain that may make use of a transportation map as a base map to show the locations of its facilities. Typically such a purchase will provide an adequate base map for years to come (Rhind, 1999).

On the other hand, organizations may wish to lease data rather than to sell it outright. This approach has the advantage of smoothing out the cash flow for government agencies. Returning to the example of the hotel chain, under a leasing arrangement, the chain would pay a regularly scheduled fee (perhaps quarterly, semiannually, or annually) for the continued use of the data (Rhind, 1999).

Another variation on leasing allows the organization that owns the information to collect royalties for its use. In this situation, the purchaser pays an agreed-upon fee at regular intervals or (more commonly) for each individual use that he or she makes of the data. For example, our hotel chain might pay a penny for each map it creates using the transportation base map.

Over time, geographic information has become more and more like a commodity, such as gold or oil or sugar. However, Cleveland (1985) notes that information has some unusual characteristics that make it different in an economic sense from the commodities and physical property with which we are familiar. Most important, the "leakiness" and "shareability" of information—especially digital information—make it difficult to control in the same ways that we control familiar property like cars or petroleum. The theft of a car or of the gas to run the vehicle deprives the owner of the vehicle or the gas to run it. The sharing or pirating of digital data does not deprive the data owner of the data, even though the pirate now has access to the data too.

Rhind (1999) makes one additional point about information as a commodity when he points out that data do not wear out. While this analysis is technically true, it is also true that data and information can become "stale." For example, day-old newspapers have little, if any, value, except to line the birdcage. More seriously, data may become obsolete, rendered useless in some circumstances, and perhaps becoming dangerous in others. For example, redrawing school district boundaries using 10-year-old data in a growing community will ignore new neighborhoods and demographic trends, making this data useless and excluding children who live in the newer areas from the redistricting plan. Similarly, using a 10-year-old database of infrastructure as the basis for "call before you dig" directions can be downright dangerous, resulting in power outages and even more serious problems.

The Role of Government in the Sale of Geographic Information

The sale of geographic data and information and other GIS products is often suggested as a "benefit" to be included in a benefit–cost analysis (see Chapter 8). As noted above, ownership of the copyright of these data sets is a prerequisite to having the right to make such sales. Different governments have different approaches to data sales. These different approaches reflect differences in the perceived role of government (Rhind, 1999).

Some governments collect information in order to support their statutory functions or otherwise fulfill their public mission, be it defense, economic development, health, environmental protection, or some other duty. In these cases, we presume that the government collected the information in order to fulfill its public mission and that it used taxpayer dollars to collect the data. We also presume that when an individual asks for public data, he or she has contributed his or her fair share and therefore is entitled to—and indeed, owns—the information. The only cost the individual must bear is the cost to reproduce the data. This is the approach of the United States, where the price of data is low.

The primary argument in favor of keeping the cost of data low is that low-cost dissemination of data maximizes the use that citizens will make of it; this is a classic free-market approach. If data are inexpensive, more people will be able to use them in the market. Therefore, people will have an incentive to develop "value-added" uses of the geographic information that was collected at public expense, and these uses will facilitate the creation of jobs and taxable wealth. If the cost of data is high, fewer people will be able to start new businesses based on data enhancement, thus inhibiting potential economic development opportunities.

On the other hand is the notion that the government is responsible for doing its job in the most efficient and cost-effective way possible. If there is some way to recover costs, or to increase revenues, then the government should do so, even if this means that the cost of data to individuals will be high. Ownership of copyrights provides an avenue through which this can occur. This position leads to the conclusion that governments should get the highest possible price for their data. This is the approach used in Great Britain. Even for academic researchers there, data costs can be high.

There are other arguments in favor of charging more than reproduction costs for public information (Rhind, 1999). One argument suggests that since only a few people have the expertise and the initiative to use public data as the raw material for economic activity, they should pay for that opportunity through higher prices that better reflect the true market value of the data for them. Another argument suggests that all citizens will share the benefit of higher priced data through the creation of an additional revenue source, and that the government should actively pursue this revenue stream. Others have argued that commercial firms are better positioned than governments to package, market, and disseminate information. Perritt (1995) adds that "if the government gives or sells public information below cost, it may undermine market opportunities for market vendors" (449–450). One final argument in favor of allowing the price of data to rise is that putting a fair market price on data improves efficiency and forces consumers of the data to specify exactly what data they need, thus avoiding "data dumps" (Rhind, 1999).

Disseminating geodatabases is fraught with legal challenges, but nonetheless is becoming more and more common, especially among government agencies. Obviously, the vast amount of data that government agencies collect makes them attractive to business and research interests alike. Government itself recognizes the economic value of selling databases, especially as other revenue streams decline at the state and local level (Onsrud, 1996). This leads into a head-on conflict between policies that encourage the sale of geographic data as a revenue source and policies that encourage or require open access to government records (Onsrud, 1999).

Embedded in this larger concern are subsidiary issues. In the United States, each state has its own laws regarding open records that influence the extent to which data can be disseminated for profit. Most states have patterned their open records laws after the federal FOIA. FOIA requires that government records be disseminated at the cost of reproduction and transfer, with no questions asked regarding the intended use of these records (Onsrud, Johnson, & Winnecki, 1996). Of course, records whose dissemination may pose a threat to national security may be withheld from use outside of its originally intended purposes. In situations where there is disagreement about whether or not specific records may be released, the legal process permits individuals or organizations to petition for the release of withheld information.

As noted at the beginning of this chapter, there are often close relationships between and among the legal issues associated with GIS. For example, legal liability associated with inaccurate public data, or the misuse of spatial data by individuals or organizations, must be addressed when organizations disseminate spatial data. Limiting secondary uses of geodatabases is one strategy that GIS database developers have embraced as a means both to "protect an agency investment in GIS or to minimize liability exposure" (Onsrud et al., 1996: 1). Secondary use of a geodatabase occurs when an individual or group who has acquired a data set from its creator disseminates it to yet another individual or group. As noted, to limit their liability, many spatial database developers include a clause in their dissemination contracts that prohibits the individual or group to whom they provide the data from further disseminating it. This restriction would not preclude a third group that wished to have the data from acquiring it directly from the developer, so the data would still be available through the original source.

Intellectual Property Rights

At odds with the concept of public access to information is the idea of intellectual property rights, which provide a legal basis for maintaining control of data, usually by the originator. Intellectual property rights as a legal concept dates back some 300 years, to British copyright law, and specifically to the Statute of Anne. The Statute of Anne (1709) was a form of state-enforced censorship, but over the years the purpose of copyright shifted to that of promoting science and the arts. An important goal of copyright is to "encourage expression of ideas in tangible form so that the ideas become accessible to and can benefit the community at large" (Onsrud & Lopez, 1998). Another argument in favor of intellectual property rights is that people should have a right to the fruits of their labor (Cho, 1998).

Traditionally, intellectual property rights have applied to several different fruits of intellectual labor: (1) original works of authorship such as *Gone with the Wind*; (2) functional inventions such as the iPod nano; (3) trademarks such as the Nike "swoop," and (4) trade secrets, such as the recipe for CocaCola (Cho, 1998; Onsrud & Lopez, 1998). Among these, it is the first, "original works of authorship," which is most applicable to GIS. Historically, original works of authorship are protected by copyright, which is covered more specifically in the section that follows this one.

Significantly, traditional criminal law related to intellectual property rights is designed to handle rights associated with tangible objects, such as the physical written word. Because of this, says Cho (1998), traditional law is ill-prepared to address issues related to the rights of those who own data and information, which are intangible property that are held on computers or in some other digital storage media, but not necessarily in a physical, written format.

Intellectual property rights have been protected internationally since 1883, when the Paris Convention established protection for industrial property. This initial treaty was instituted by an agreement among 14 member states after international exhibitors "refused to attend the International Exhibition of Inventions in Vienna in 1873 because they were afraid their ideas would be stolen and exploited" outside their home countries (World International Property Organization, 2006: 4). In addition to establishing this initial treaty, the Paris Convention also put in place an international bureau to organize subsequent meetings and handle other ongoing administrative tasks.

Later treaties of this group added protections for literary and artistic works (Berne Convention), trademarks, patents, and other properties. The cornerstone of international intellectual property conventions are the Paris Convention and the Berne Convention (World International Property Organization, 2006). By 1893, the two small bureaus of the Berne and Paris Conventions joined together to form the Bureaux for the Protection of Intellectual Property, which has evolved into the World Intellectual Property Organization (WIPO). In 1974, WIPO came under the auspices of the United Nations. Today, more than 180 countries, comprising more than 90% of world states, are members of WIPO (WIPO, 2006).

WIPO is well aware of the need to address the unique concerns about digital intellectual property and has established a "Digital Agenda" to respond to "the confluence of the Internet, digital technologies and the intellectual property system" (WIPO, 2006). An important part of this effort is to both define and harmonize intellectual property protection among the member countries.

Enforcement of intellectual property rights internationally is weak, relying primarily on the willingness and initiative of member countries to comply with the "best practices and procedures" of WIPO. Less developed countries, developing countries, and some others are sometimes lax in their enforcement, in part because of inadequate resources to devote to it (WIPO, 2006). Notoriously, China is among the countries where enforcement has been lax and pirating (especially of movies and DVDs) has been rampant.

The next section discusses the use of copyright to protect intellectual property rights associated with GIS projects, especially data.

Copyright

Copyright is the legal means to protect intellectual property rights of original works of authorship, including GIS data and products (Onsrud & Lopez, 1998). It is the "exclusive right given to a creator to reproduce, publish, perform, broadcast and adapt" a work and is designed to prevent unauthorized copying (Cho, 1998: 176). Corey (1998) describes copyright as a tool "to control and extract a return for the use of intellectual property" (39). As noted in one of the preceding sections, ownership of copyright is necessary for the sale of geographic data or information.

Copyright law, which today has the connotation of prohibiting anyone other than an author from copying a written work, originated as a right to copy during the Middle Ages. In those days, it was understood that the author possessed such material, and was solely entitled to give permission to others to (literally) copy the manuscript by hand. Furthermore, the author/owner was entitled to charge a fee for the copyright if he or she desired. Religious monasteries, which were among the most prolific producers of writing, sometimes charged such a fee (Cho, 1998: 190).

From these humble beginnings grew the legal concept of copyright. Under the terms of the Berne Convention (as mentioned in the preceding section) copyright protection automatically resides in original works of authorship, with no mark or notice required. Moreover, publication is not a prerequisite for copyright protection (Harris, 1998; Onsrud, 2005).

However, in order to qualify for copyright protection, the work "must be fixed in a tangible medium of expression" (Onsrud, 2005) and be capable of identification (Harris, 1991). Paper maps were first covered by copyright protection as a result of the Berne Convention of 1886, under the rubric of "literary and artistic works" (Harris, 1991). Today, digital mapping media also meet this standard (Cho, 1998; Harris, 1998). According to Onsrud (2005), GIS data sets and software both meet the requirement and qualify as works that may be copyrighted.

Over the years, copyright law has evolved to comprise several key elements. In general, the creator of the material owns the property. The major exception to this rule of thumb is that when a person creates such material as part of his or her employment, then the employer owns the copyright. However, even when this situation holds, the moral rights remain with the author (Harris, 1991).

Cho (1998: 191) reports that the publication of a work automatically triggers copyright protection. Registration of a work is not required by the Berne Convention in order to establish copyright protection; it is recommended, but not required, that the creator of a map insert a copyright notice in order to alert users that a copyright on the material exists (Harris, 1991). Furthermore, a copyright is valid as long as its last author survives (Cho, 1998; Onsrud, 2005), and remains valid for an additional 70 years in the United States (Onsrud, 2005). To qualify for copyright protection, a work must be both original and the product of an independent intellectual effort (Cho, 1998; Harris, 1991; Onsrud, 2005). Harris notes that if a work is "made for hire," then the copyright on it lasts for 75 years from its date of publication, or for 100 years from the date it was created, whichever comes first (Harris, 1991).

As previously noted, the U.S. federal government by law cannot copyright data that its collects. Harris (1991) emphasizes the importance of international copyright treaties such as the Berne Convention, which fall under the jurisdiction of WIPO. Treaties such as these "impose the same level of minimum protection in countries around the world" (Harris, 1991: 72) and create a global network of copyright relations. Moreover, as more countries join the convention, the treaties become more valuable (Harris, 1991: 72).

A GIS data developer may find him- or herself in the position of either the creator of an original database or as the user of another copyrightable database. Many databases today are available at no charge, but it is still necessary to give credit to the owner of the database or other material.

Data Privacy

Data privacy and confidentiality are a growing concern in any discussion of computer technology. The ability of computers to store large quantities of personal information and to combine a variety of public and private data sets joins with the "leakiness" of information (Cleveland, 1985) to warrant a high level of concern. The ability of GIS to aggregate and disaggregate information, along with their spatial search and overlay operations which facilitate matching personal data, have added potential to erode individual privacy, or as Dobson (2003) might put it, to foster "geoslavery." Moreover, the explosion in the collection of data by public and private entities, coupled with an increase in clandestine data monitoring by the U.S. federal government in the post-9/11 era, raise new questions about data privacy for individuals. Admissions in late spring 2006 that a laptop containing personal information (including Social Security numbers) of U.S. military veterans and service members was taken home by an employee of the U.S. Veterans Administration and subsequently stolen from the employee's home demonstrates that privacy concerns are worthy of attention. In addition, identity theft has become both a significant crime and a major worry for citizens in this increasingly digital age. Data privacy is no longer a hypothetical concern; it is a very real threat.

In addition to concerns among the general public, privacy and confidentiality are becoming increasingly important concerns among GIS developers and users. What makes these issues so important is the ability of GIS to combine previously unrelated databases with specific locational information, thus providing new means by which we can track the whereabouts of individuals. The ability to match, for example, an insurance company's database of policyholders with a cigarette manufacturer's database on smokers could result in an individual's losing his or her policy or being charged a higher cost to take into account the potential health hazards of smoking.

Even more troubling, however, are the potential consequences of mismatching databases. For example, Palast (2001) describes how election officials in Florida contracted with Choicepoint, a private data analysis group, to cross-match a list of convicted felons against voter registration lists. Choicepoint itself acknowledged that the list included a large percentage of false positives—in the range of 90%. Florida election officials gave these error-filled lists of names to poll workers in Florida, who then prevented many legally eligible voters from casting their ballots. As "luck" would have it, the majority of the legal voters who were prevented from fulfilling their civic duty were African American.

Crampton (2006) notes another venue where false positives in matching databases can have troubling outcomes. Speaking about the rise of geosurveillance since September 11, 2001, to identify people who are "dangerous," Crampton points out that the problem of false positives in the search for potential terrorists means that a large number of innocent people are likely to face arrest in the search for the guilty.

Any discussion of data privacy and confidentiality must acknowledge that both public and private organizations maintain large databases containing sensitive material. Most often, concerns about privacy and confidentiality stress the dangers of government-held data, citing Orwellian "Big Brother" scenarios. Jerome Dobson (2003) describes the increasing use of geographic information to identify us not only by name, but also by location, as "geoslavery." Onsrud, Johnson, and Lopez (1994) specifically discuss this cozy relationship between "government and commercial sector 'insider elites' that are compiling and using expansive knowledge about individuals' lives" (1084). They note that "when asked, most people are unwilling to have personal information about themselves passed on to others for non-specific commercial or government purposes" (1084).

Our governments (local, county, state, federal) collect and maintain vast amounts of personal information about us. The various governmental jurisdictions record our births, our marriages, our finances (in great detail), our property ownership, our employment history, our military service (or lack thereof), the make and model of our cars, our brushes with the law, our deaths, and many other details. Because much of this information is in the public record, it is readily available to anyone and everyone who wishes to view it. Historically, our spatial data privacy has hung on a single thread: that all this information has been housed in a large number of separate agencies and private organizations that interact on a limited basis.

That this information has historically been kept in hard-copy, non-digital format has made it difficult to integrate these disparate data sets. While it has been difficult in the past, it has not been impossible. Indeed, private companies have often made use of these data sets to promote their business interests. For example, companies regularly monitor birth records and collect and maintain this information. This enables them to send just the right coupons for just the right products at just the right time to just the right potential purchasers. The coupons for formula arrive shortly after the baby's birth; the coupons for "pull-up" type diapers arrive when the child is nearly 2; the coupons for books based on the "Barbie" character arrive when the child is between 4 and 5 years of age—but only if the child is a girl.

Today, the ability to match and integrate these separate and disparate digital databases collected by the government poses a severe threat to our privacy, especially when they can be integrated with and augmented by private data sets in a for-profit setting. Increasingly, especially in the post-9/11 era, these formerly discrete data sets have been combined under the premise that doing so will enhance national security and protect us from "bad guys." GIS provides a far more efficient way to bring together these disparate databases, while adding a geographic location to information.

Moreover, since the terrorist attacks in the United States in September 2001, the U.S. federal government has increased surveillance of residents of the United States. This increased surveillance includes wiretaps and other forms of surveillance that previously required permission from the courts. Using the September 11 terrorist attacks as justification, much of the additional surveillance has been authorized by executive order of the president. Now both the means and the motivation for additional losses of data privacy are in place.

As previously noted, private entities also house huge databases. For example, both the *Wall Street Journal* and *BusinessWeek* reported that Philip Morris, the U.S. cigarette manufacturer, has compiled a database that includes the names of 26 million smokers—about one-tenth of the U.S. population. The company built this database through perfectly legal means: a premium give-away program that allowed people to redeem empty cigarette packages for consumer goods such as hats and T-shirts (Berry, 1994; Shapiro, 1993). In 2005, *BusinessWeek* refers again to Philip Morris's "growing database of 26 million smokers to whom it sends everything from birthday coupons to the chance to attend events like November's birthday concert" (Byrnes, 2005).

Such a database would certainly be of interest to insurance companies that charge higher premiums to smokers. Potential employers who prefer to hire nonsmokers would also find this database useful. Philip Morris itself uses the database to send "money-off" coupons to people whose names are included, possibly derailing the efforts of smokers who are trying to quit. A company spokesperson says that the database is far too valuable to consider selling it or leasing it. Still, all it takes is a change in company policy to breach the privacy of the 26 million people whose names are in the Philip Morris database (Shapiro, 1993).

Credit reporting bureaus are another private business that maintain huge databases of sensitive information. Recent reports in the United States indicate that a surprisingly large percentage of this information is erroneous. Still, this information is readily accessible. Banks often make use of these files to identify individuals who could qualify for one of their bank credit cards. In such cases, the credit bureaus release the information with neither the knowledge nor the approval of the individual to whom the record belongs! In some instances, individuals have been rejected for credit because of too many requests for their credit history as banks review individual credit records in their search for potential new customers. Increasing numbers of employers also run credit checks on prospective employees.

In addition to databases, cameras—either hidden or open to public view—also track our movements. Today, most retail stores have closed-circuit cameras recording transactions. Increasingly, public streets and public transportation are also under constant surveillance via closed-circuit TV. Monmonier's book *Spying with Maps* (2004) provides an excellent discussion of the many devices used to track our location.

Law regarding data privacy is undergoing serious revision these days. The trend appears to be toward diminution of privacy rights. In general, legal decisions regarding Fourth Amendment protections against unreasonable search and seizure have held that there is a difference between what occurs inside our homes (where we may expect greater protection of our privacy) and what occurs in the public space (Herbert, 2006: 5). For example, a 1983 ruling in *United States v. Knotts* (460 U.S. 276, 1983) held that police need not obtain a warrant before using a radio beeper to monitor the location and movement of a vehicle on public thoroughfares. The logic is that the radio beeper is merely an extension of a police officer's ability and authority to monitor the vehicle visually (Herbert, 2006: 5).

On the other hand, the use of a thermal scanning device without warrant to monitor unusually high release of heat from a private residence where the owner was suspected of using high-density halide lamps to raise marijuana was ruled a violation of the Fourth Amendment (*Kyllo v. United States* 2001; 533 U.S. at 40). The logic of this decision is that thermal scanning devices are specialized equipment that is not used by the general public and that their use was not merely an extension of an ordinary and usual physical ability but rather an entirely new way of seeing into a person's private residence. Therefore, use of this technology requires a warrant (Herbert, 2006: 7).

The prospects for increased protection of data privacy through government regulation appears limited (Onsrud, Johnson, & Lopez, 1994). In the United States, 1973 regulations by the U.S. Department of Health, Education and Welfare (HEW) were incorporated into the Privacy Act of 1974. These regulations contain five privacy protection principles:

1. There must be no secret personal data recording systems.
2. Individuals must have a means of learning about their stored personal information, and how it is used.
3. Consent should be required for secondary uses.
4. Individuals must have a means of learning about their stored personal information.
5. Data controllers must maintain and ensure data security. (Onsrud et al., 1994)

Internationally, the 24 leading industrialized countries (including the United States) of the Organisation for Economic Cooperation and Development (OECD) adopted a set of privacy guidelines in 1980 (Onsrud et al., 1994). OECD guidelines have eight principles:

1. There should be limits to the collection of personal information. Collection should be lawful, fair, and with the knowledge and consent of the individual.
2. Data should be relevant, accurate, complete, and up-to-date.
3. The purpose of the data should be stated upon collection, and subsequent uses should be limited to those purposes.
4. There should not be any secondary uses of personal information without the consent of the data subject or by the positive authorization of law.
5. Personal data should be reasonably protected by the data collector.
6. Developments, practices, and policies with respect to personal data should follow a general policy of openness.
7. Data subjects should be allowed to determine the existence of data files on themselves and be able to inspect and correct data.
8. Data controllers, whether in the public or private sectors, should be held accountable for complying with guidelines. (Onsrud et al., 1994)

Other protections for data privacy come through codes of ethics. Organizations adopting relevant codes of ethics include the Association of Computing Machinery (ACM), the Direct Marketing Association, the Information Industry Association, and the National Information Infrastructure (Onsrud et al., 1994). In addition, the GIS Certification Institute (GISCI) is currently developing a code of ethics and procedures to address violations of the code by its members (*www.gisci.org*). Keep in mind that codes of ethics do not have the weight of law, even though they are a step in the right direction.

This section is intended to provide a helpful introduction to the legal framework of data privacy, rather than to be exhaustive. For organizations that collect data for inclusion in their GIS, it is wise to review both current law (which is always subject to change) and codes of ethics of relevant organizations in order to identify a standard of care with respect to collection, maintenance, and use of data files that may compromise individual privacy rights. The concepts, cases, and questions covered in this section are intended to point you in the right direction.

Evidentiary Admissibility of GIS Products

Another important question related to the law and GIS is how GIS fares as evidence in courts of law. Cho (1998: 232) identifies a key quandary in this discussion: on the one hand, if GIS is considered purely as a representation of facts, then its admissability is assured; on the other hand, because GIS is considered as a literary or artistic work under copyright laws, then problems arise. Simply put, an artistic or literary work is by definition not factual (Cho, 1998). An additional complication to evidentiary admissibility of GIS products arises because of the ease with which digital data may be altered, leaving no evidence of any changes (Onsrud, 1992). Thus, the path to the use of GIS as evidence in a court of law has not been smooth.

Evidence is material offered in a court of law in order to persuade a "trier of fact" (i.e., a jury) about the truth or falsity of a disputed fact (Cho, 1998; Onsrud, 1992). Onsrud (1992) notes that all federal courts in the United States abide by the rules of evidentiary admissibility published in the *Federal Rules of Evidence* and that states follow similar sets of published rules. Moreover, notes Onsrud (1992: 6) higher courts nearly always uphold the use of evidence allowed in lower courts, unless there is a "clear abuse of discretion."

The first and most important hurdle that evidence must clear in order to be deemed admissible is that of relevancy. In order to be relevant, it must have "probative value. To have probative value means that it can go to prove a material fact" (Cho, 1998: 235). Ordinarily, demonstrating the relevancy of GIS evidence is not difficult (Onsrud, 1992).

On the other hand, there is more to evidentiary admissibility than relevance, and these are more difficult challenges. Because of the difficulties associated with GIS data described above (GIS is copyrighted as literature or art, and the ease with which digital GIS data may be altered), GIS evidence is treated as "hearsay" (Cho, 1998; Onsrud, 1992). Hearsay is an "oral or written assertion . . . other than one made by the declarant while testifying at the trial or hearing, offered in evidence to prove the truth of the matter asserted" (Fed. R. Evid. 801 [a] and [c], as reported in Onsrud, 1992). In general, hearsay is not admissible as evidence, except in a few, well-defined instances, and in fact, the "hearsay rule" is often used to exclude computer-generated exhibits as evidence (Onsrud, 1992).

Computer information—including GIS—is considered to be hearsay because it is developed outside of the courtroom, not on the witness stand. If a witness in a court of law had "designed and manufactured the computer hardware, wrote the GIS software, and carried out the product generation or database manipulation procedures involved in the dispute," then the GIS could be considered as evidence other than hearsay (Onsrud, 1992). Of course, this is an impossible hurdle to clear. As a result, any computer- or GIS-based information may be considered admissible as evidence only if it qualifies as an "exception to hearsay rules" (Cho, 1998; Onsrud, 1992).

There are several categories of exception to hearsay rules. The first of these is the "business records exception" (Cho, 1998; Onsrud, 1992). In order to qualify as a business record, the GIS product must be developed as a regular business practice of its creator. Under this rubric, the GIS records of a local government would most likely be admissible, while a GIS developed by a community group for a one-time local political battle would probably not qualify (Onsrud, 1992).

In order to meet the standard of admissibility under the exception to hearsay rules, digital data or GIS products must also be determined to be "authentic" (Onsrud, 1992). There are three steps to providing such authentication: (1) show the input procedures used to get the data into the computer; (2) show which tests were used to ensure the accuracy and reliability of both the computer operations and the information that produced the data; and (3) demonstrate that the computer record was generated in the regular course of business, and that the business did indeed rely on it (Onsrud, 1992).

In practice, the courts also often require that an "authenticating witness" verify the database on the witness stand where he or she will face cross-examination (Onsrud, 1992). An authenticating witness must therefore be familiar with all aspects of the project, including procedures in the field and in the office. This witness must be able to explain persuasively why errors are unlikely to be present in the GIS operation, as a means to document the validity of the data and ensure its admissibility as evidence (Onsrud, 1992). This can be a daunting task that constitutes a significant burden for anyone trying to admit computer records into evidence. As a result, some courts have eased this burden and asked only that the custodian of the data testify that the records have, indeed, been maintained as a regular business practice, unless there is evidence that the data are not trustworthy (Onsrud, 1992).

In at least three states (Iowa, Virginia, and Florida), state legislatures have passed laws stating that computer records are held to be admissible as long as they are collected and maintained in the ordinary course of business (Onsrud, 1992).

In addition to the "business records" exception to the hearsay rule, there is also an exception made for public records and reports (Onsrud, 1992). In these cases, the custodian of the data must certify that the printout or other material is correct and accurate (Onsrud, 1992).

There is one other way that GIS data may be introduced into the courtroom. This is as "demonstrative evidence" (Cho, 1998; Onsrud, 1992). Demonstrative evidence is tangible and is used in the courtroom to help jurors understand complex issues and situations, but it is not "real" evidence because it has no probative force (Cho, 1998; Onsrud, 1992). Because of this, notes Cho, demonstrative evidence is not allowed in the jury room during deliberations, as real evidence would be.

Cho notes one other way that GIS materials may be introduced in the courtroom. This is as "novel scientific evidence" (Cho, 1998: 246). This avenue is not available outside the United States. To be admissible, the material must ". . . have gained general acceptance in the particular field in which it belongs" (Cho, 1998: 246).

Onsrud notes an interesting contradiction regarding GIS evidence. On the one hand, the courts generally have made it difficult for GIS materials to be admitted as evidence. On the other hand, the visualization capabilities of GIS evidence, like that of photographs, have made it highly persuasive to those who view it. This must all be leavened by the knowledge that today's computer technologies make it easier for the average person to alter graphic and other visual information with the click of a mouse button (Onsrud, 1992).

Conclusions

Legal issues in GIS have evolved dramatically over the past decade. In addition to encompassing a growing number of concerns, the concerns are increasing in complexity and interrelatedness. One issue that is likely to grow in significance is database security and protection.

It is all but certain that law regarding GIS will continue to evolve in the coming years, encompassing both case law and public statute. It is critical that GIS professionals consult applicable statutes and case law in order to ensure that their operations are in compliance with all applicable laws.

Chapter 14

Ethics for the GIS Professional

As the GIS community has evolved from a loose coalition of practitioners of geographic information science into a genuine profession (as documented in Chapter 12), it has adopted a key element of a profession: a code of ethics. This has not happened overnight, but rather represents a decade-long effort led in large part by Will Craig (1993, 2004). And while the GIS profession has made a great deal of progress in working to foster ethical behavior among its members, there is much left to do.

There are two major elements to ethics for the GIS professional. Ethics begins with competency, the ability of an individual to carry out the performance of his or her professional duties accurately and correctly. Ethics extends to include a moral obligation to behave in a moral manner in all professional activities.

This chapter focuses on ethics among GIS practitioners. It begins with a discussion of the history and sources of concern about ethics in GIS. This section first explains the evolution of GIS as a precursor to the need for ethics, then describes the development of discussion of ethics in the field. The following section provides a general definition of ethics and discusses five important relationships where ethics are important to the GIS professional. This part includes a discussion of the ways in which specific ethical issues have arisen among GIS professionals. The chapter concludes by presenting the GIS Code of Ethics established by the GIS Certification Institute, and then details future activity needed to foster ethical behavior among GIS professionals.

Portions of this chapter are from Obermeyer (1998). Copyright 1998 by Elsevier Science Ltd. Adapted by permission.

History of the GIS Ethics Movement

The need for a GIS code of ethics rose in tandem with the growth and diffusion of GIS technology and the rising number of people in the field. Throughout much of the evolution of GIS, the technology itself filled a small niche, and the people who worked with the technology were small in number. In the early days, the research and programming that underpinned the technology promised more hard work than profits.

In addition to providing the research needed to make GIS a reality, the founders and early developers of the technology also served as mentors and teachers, expanding the number of people working in GIS. From these founders, their protégés and students not only learned the theory and techniques of GIS, but also became socialized into the ethics of the fledgling profession, and learned through example the importance of professional integrity.

Well into the 1980s, it seemed as though virtually everyone in the GIS community knew everyone else. Newcomers frequently had the opportunity to meet and often to work with the legendary developers of GIS and get to know them, even if just a bit. GIS developers who would qualify for the "GIS Hall of Fame" (if there were one) were unusually accessible and encouraging, gauging by the experience of at least one of the authors of this book.

In short, the degrees of separation between founders and newcomers were very few in those bygone times. When you asked someone about another person in the GIS community, chances are they themselves either knew the person in question or knew someone who knew him or her. Informal links, including personal and professional friendships and the inevitable related word-of-mouth were extremely important mechanisms for evaluating GIS practitioners and spreading the word about their level of competency and integrity.

It is important to keep in mind that in the early days of GIS, diffusion of the technology itself was at a very different stage too (Dobson, 1993). Most GIS were specially built from scratch by a team of GIS professionals who worked closely with software developers. This was essential because there were no "off-the-shelf" GIS available. In terms of the product development, early GIS were more or less "Beta" implementations. The people building early GIS and its related infrastructure were working as much to prove the technology as they were to develop a GIS for a specific application. Implementation by implementation, GIS improved, until it was more widely available, in off-the-shelf packaging, for a more general audience.

As GIS technology matured and found a wider audience, the profits associated with it became an attraction. GIS has become an industry. The industry includes the obvious elements: software development, consulting firms, GIS professionals. But the industry also includes educators, trainers, and an increasingly vast array of ways to learn GIS (surprise, surprise: education itself has also been transformed by the advance of technology).

When the number of people associated with GIS grew, so did the degrees of separation between founders and newcomers. No longer could you count on asking your buddies about an individual and getting an answer, or at least getting a referral to someone else as before. More and more people were attracted to the field because of the good career potential. Most of them were pretty much unknown. The expansion of GIS training programs to include online education made it even more difficult to trace professional pedigrees and tell who was who. It is this diffusion of GIS technology, accompanied by a rapid expansion of the GIS community, that has made GIS ethics an issue of growing concern.

And when the number and variety of GIS applications grew, so too did the potential for ethical lapses to cause problems. Whereas early GIS applications were closely monitored by a relatively small cadre of professionals, today anyone with a credit line of $40 and access to Amazon.com can purchase a fully-functioning GIS with the "Getting to Know ArcView GIS" tutorial and get to work, with or without adequate knowledge or awareness of the pitfalls. And while the software turns into a pumpkin after the 6-month trial period, this is all it takes to get started.

It was in the late 1980s that ethics first showed up on the radar screen of the GIS community. Craig (2004) identifies the 1986 meeting of the Urban and Regional Information Systems Association (URISA) as the seminal moment in the development of GIS ethics. Specifically, he cites the central message of Marshall Kaplan's keynote address as the simple but profound idea that stimulated his passion for GIS ethics: "Consider the impact of your work" (Craig, 2004). Craig used this message as a motivating force, and began to gather codes of ethics from a large number of professional organizations. He reviewed these codes over a period of several years, and became the most influential champion of a code of ethics for GIS professionals.

Researchers on GIS implementation have been exploring the issues surrounding professional competence, responsibility, and ethics in GIS since the early 1990s (see, e.g., Craig, 1993). One of the first groups in the

GIS community to address the issue of competence (the most basic aspect of ethics) was the professional organization of land surveyors, who proposed instituting a licensing program for GIS professionals on a state-by-state basis. This initiative failed to gain the support needed to institute a licensing program throughout the United States, but the issue of ensuring the qualifications of GIS practitioners has remained on the radar screen ever since (Craig, 1993; Obermeyer, 1993).

Concerns did not end there, however. GIS trade publications raised awareness about ethical lapses in the business end of GIS technology (Linden, 1991). And it was not long before critical theorists in the field of geography began to express concerns about ethical issues at the societal level related to implementation of GIS (e.g., Curry, 1995; Monmonier, 1998; Pickles, 1995; Rundstrom, 1995; Sheppard, 1995; Yapa, 1998).

A major impediment to establishing a code of ethics for GIS practitioners was the absence of a single, unified professional organization for the multitude of people working in the field. Craig (1993) had suggested that the consortia of organizations that historically sponsored GIS meetings such as GIS/LIS take joint responsibility for developing and implementing a code of ethics for GIS specialists. Another group that had been a prospect in this role is the University Consortium for Geographic Information Science (UCGIS) (Obermeyer, 1998); in fact, UCGIS has stepped up to the plate in the area of GIS education. Ultimately, it was the Urban and Regional Information Systems Association that took the lead on developing a U.S. national GIS certification program that essentially completed the evolution of the GIS community into a profession.

Over the years, the ethics discussion continued, absorbed, in part, into the larger debate on certification (discussed in Chapter 12 of this book). As the GIS community evolved into a profession, and that profession developed a national certification process, the development of a code of ethics remained a critical part of this process. As at the beginning, Will Craig continued to champion this cause, and ultimately authored the GIS Certification Institute's Code of Ethics, which became a reality in 2003.

Ethics Defined

A general definition of ethics may be found in any standard dictionary. The *American Heritage Collegiate Dictionary* defines *ethics* as "a set of principles of right conduct; a theory or system of moral values." The website for

the Joseph and Edna Josephson Institute of Ethics in Los Angeles provides a more thorough definition:

> Ethics refers to principles that define behavior as right, good and proper.
> Such principles do not always dictate a single "moral" course of action, but
> provide a means of evaluating and deciding among the competing options.
> (Josephson, 2005)

The site, under the director of Michael Josephson, goes on to point
out that "ethics" and "values" are not synonyms. Whereas ethics are
concerned with how a moral person *should* behave, values are the inner
judgments that determine how a person actually *does* behave. Furthermore, Josephson believes that values concern ethics *only when* values pertain to beliefs about what is right and wrong. For example,
the desire for wealth is a value, but certainly not an ethical value
(Josephson, 2005).

Onsrud (1995) establishes a similar contrast between legality and ethics. He begins by defining ethics as "behavior desired by society that is
above and beyond the minimum standards of behavior established by law"
(90). He goes on to discuss the relationship between law and ethics, noting that some behavior is both unethical and illegal (e.g., murder and
extortion). Conversely, most ethical conduct is also legal, although there
are exceptions. For example, a father driving 60 miles an hour in a 40
mph zone on a deserted stretch of highway to transport his seriously ill
child to the hospital is violating the law to save his child. His actions, while
clearly in violation of the law, are not unethical. Then again, some behavior that most societies would classify as unethical is not illegal. For example, a student caught with an illicit copy of tomorrow's exam is cheating,
and may be held accountable by the educational system where she is
matriculating, but she is not subject to legal sanctions by the state (even
though many instructors might wish otherwise).

There are other definitions of ethics that specifically refer to GIS.
For example, Crampton (1995) points out that in GIS and cartography, ethical behavior had come to be equated with good professional
conduct—for example, adhering to accuracy standards. Curry (1995)
agrees, and goes on to propose that GIS professionals must accept their
responsibility to meet such standards. Moreover, says Curry, cartography
and GIS exist within the larger context of the system of science which
"operates on the assumption that it is at its heart a moral system, that its
practitioners can be counted on to engage in a set of practices that exem-

plify a set of values—honesty, altruism, communalism and disinterestedness" (Curry, 1995: 60). In this context, impartiality is a synonym for disinterestedness; this is also what Weber (1946) describes as "without regard for persons."

Curry's discussion is important because it brings out an important point that is a theme running through this chapter: that professions fulfill an important role in imparting a culture of ethical behavior to their members. Similarly, Ehrenreich (1990: 139) has pointed out that "nonviolent social control" was an important justification for many professions in the United States. Obermeyer (1998: 220) emphasizes this point, drawing from the work of Pugh (1989) and Weber (1946):

> An extremely important role that professional organizations play . . . is to help shape professional culture and convey it to incoming practitioners. This professional culture includes norms and conventions of ethical behavior.

When considering responsibility and ethics in GIS, it is important to remember that there are many links in the chain the produces GIS products, including hardware, software, and data (Curry, 1995). Each of these areas of responsibility is typically associated with a specific individual or group of people. Moreover, the map user him- or herself must also assume some responsibility as the "driver" of the GIS vehicle (Curry, 1995: 62). This view is at odds with Gersmehl's, as recounted in his 1985 parable "The Data, the Reader, and the Innocent Bystander." Recounting his own experience as a cartographer whose map a reader in the U.S. Department of Energy misinterpreted and consequently misused, Gersmehl counseled other mapmakers to try to anticipate all the possible ways that their maps could be misused. In so doing, he generously (probably too generously) let map users off the hook. In reality, ethical behavior in cartography and GIS is the individual responsibility of each person with a role in the creation, implementation, development, analysis, and use of maps and other GIS products.

Ethics and Professional Obligations

Embedded in the very idea of ethics is the idea that Marshall Kaplan raised in his 1986 keynote address to the Urban and Regional Information Systems Association: "Consider the impact of your work" (Perkins, 2005). This simple yet profound statement implies an outward-looking obligation. How do my actions affect others? This brings us to an important question: To whom do professionals (especially GIS professionals) have ethical obligations?

Craig (2004) and Obermeyer (1998) are largely in agreement on the groups to whom GIS professionals have ethical obligations. Both identify (1) professional colleagues and the profession; (2) employers, clients, and other funders; and (3) society among the key groups to whom GIS professionals have professional obligations. But Craig adds (4) individuals at large, while Obermeyer adds (5) students. In this section, we discuss the nature of the responsibility that GIS professionals have to each of these groups.

Responsibility to Professional Colleagues and the Profession

As we start with the inner circle, GIS professionals have a responsibility to professional colleagues and the profession as a whole. As previously noted, an important purpose of professions is to instill and foster a sense of ethics among its members. But, as Obermeyer (1998: 222) puts it, "instilling a sense of ethics is the last step of a process that begins with good practice."

The central role of a profession is to set a basic standard of competence. This standard may be explicitly set and formally enforced, as when a profession adopts a formal licensing procedure supported by governmental regulations (usually set at the state or national level). The legal, medical, and surveying professions all employ this strategy. Under the terms of compulsory licensure, individuals are permitted to practice their profession if and only if they successfully negotiate all the rites of passage (Obermeyer, 1998).

Some professions explicitly set standards of good practice (competence) but do not formally enforce them. A case in point is the American Institute of Certified Planners (AICP), which has a voluntary certification program. Since the AICP has not taken the additional step of enlisting the support of government agencies to require licensure based on its standards, most jobs in planning are not reserved for certified AICP professionals. The certification program of the GIS Certification Institute, like that of the AICP, is also voluntary.

Several authors have observed that within the GIS community, professional obligations to colleagues and the profession as a whole seem to be the most well developed, whereas ethics directed more toward the greater society are in the greatest need of development (Crampton, 1995; Curry, 1995). Obligations toward professional colleagues and the GIS profession tend to emphasize "good practice," and have more or less focused on the need for basic (which is not necessarily minimal) competence.

Obligations to Employers, Clients, and Other Funders

A second ethical obligation is the one between the professional and employers, clients, and other funders. In this relationship, the concept of "good practice" rises to a higher level. Honesty is at the core of this obligation, and requires that the GIS professional must establish and maintain an honest relationship with employers, clients, and others with whom they have professional relationships. Linden (1991) suggests that some suppliers of GIS goods and services are not always living up to their ethical obligations.

As has been noted repeatedly in this book, one of the consequences of the growth of GIS has been the commensurate increase in the value of the GIS market. This has resulted in growth in the number of vendors of hardware, software, and services competing to serve this lucrative market. Linden (1991) makes the point that many of the people who have been attracted to the GIS industry as it has matured have as their primary interest profit. He contrasts this group with the earlier members of the GIS community, who shared an interest in inventing and nurturing the technology that bound them together. We may infer from Linden's discussion that the growing membership in the GIS community has made it much harder for peer pressure alone to show newcomers the right path and to keep them on it. Nor should we be surprised that the lure of profits has proven too enticing for some, resulting in some unethical marketing practices (Linden, 1991).

The nature of these ethical lapses include exaggeration of system/hardware performance; spreading rumors about competitors' products and services; intentionally underbidding contracts, then claiming a misunderstanding in order to justify a later price increase; and embellishing or otherwise misrepresenting company credentials (e.g., making a short-term assignment sound like a lifetime experience) (Linden, 1991). Ethical lapses such as these drive up the costs for GIS users, damage the hard-earned reputations of the many honest vendors, and have the potential to harm the profession itself over the long term (think, for example, of the common perception of used car salespeople or lawyers) (Obermeyer, 1998).

The free-market response to this problem would be to assume that over time, the dishonest element will eventually lose market share and ultimately be forced out of business. Realistically, this laissez-faire approach ignores the fact that other dishonest vendors may come to take the place of those who have fallen by the wayside before them. In fact, they may not have a long-term interest in the profession, but prefer to make a quick buck. The free-market approach also fails to acknowledge the great harm that may be done to an individual GIS user as he or she waits for market forces to drive the riff-raff out of business. Thus, the concern for integrity is important, and may be especially so for first-time GIS implementors, who may be unsure about which system to adopt and how to achieve the organizational expertise they will need to make best use of GIS technology (Obermeyer, 1998).

Obligations to Society

Direct users of GIS are not the only ones who have a stake in the technology. GIS (and other information technologies) have become integral tools in both the public and private sectors. As Obermeyer (1998: 223) puts it, the fact

> that each and every member of society is a participant, if not as an informed [GIS] user, then as an unaware data point. From birth to death, we become part of public records. In the United States, for example, one of the first tasks of new parents is to apply for their baby's social security number, which will stay with her throughout her lifetime. And it is not only public agencies that record a child's birth. The perfectly timed arrival in the mail of free samples of and coupons for commercial products that are developmentally appropriate for the child as she grows provides clear evidence that the child has become a data-point in private data-bases as well.

People have long expressed concern about the relationship between people and technology. In recent years, this issue has received extensive attention as it relates to the proliferation of GIS and cognate technologies. Monmonier's *Spying with Maps* (2002) provides an insight into locational surveillance that runs the gamut from amusing to chilling. Gutmann (2004) describes how several U.S. IT firms are developing technology that makes it easier to spy on Chinese citizens in virtual space. Even today's cell phones are equipped with GPS to keep track of our movements.

The key to understanding the importance of the obligation of GIS professionals to society is to acknowledge that GIS is not just a

> tool designed to solve one aspect of a particular problem—that of translating spatially referenced empirical information into a spatial language to enable cartographic representation of patterns and relationships, and of analyzing the nature of these relationships; rather, the development of GIS, or any other technology is a social process. (Sheppard, 1995: 6)

GIS professionals should bear in mind that the inventors and developers of GIS made conscious and deliberate choices about the configuration of the technology based on contemporary societal and technical conditions. These GIS developers worked within specific institutional environments (largely white males working in institutions of higher education in North American and Europe) that specified the boundaries of their task. Moreover, existing technology, software logic, and specific spatial theories also influenced their choices as they worked. These, in turn, limited the kind of GIS that are available today.

A number of scholars (Aitken & Michel, 1995; Curry, 1995; Obermeyer, 1995a; Rundstrom, 1995; Weiner et al., 1995) have pointed out that one disturbing result of this process is that many groups have been poorly represented in today's GIS. This theme is echoed by the work of the National Research Council (2003). The use of GIS and other information technologies can make it harder for the average citizen to participate in ongoing policy debates. This occurs because using GIS employs new techniques, analyses, and graphical materials that lend a new aura of persuasiveness to policy reports prepared by public and private organizations. No matter how sound or unsound the underlying ideas, the GIS can make a report seem more authentic and authoritative than it might otherwise appear. As Monmonier notes, "The map is a robust medium, and even bad maps may communicate" (1993: 3). This leaves individuals and citizens' groups at a disadvantage if they do not also have access to GIS, and they may find it difficult to mount a successful challenge to the powers that be.

There are examples where GIS has been put to use in order to challenge the status quo. For example, a Minnesota state legislator used a desktop demographic program, Census Bureau population statistics, and state budget data to create his own map that showed that a small group of wealthy suburbs had only 25% of the area's population, yet received 70% of new jobs, 83% of highway funds, and 90% of new sewer funding. As McNulty (1994) described it, this information "set his constituents rocking."

On another front, Smith (1992) discusses the prominent role that GIS and GPS played in the first Gulf War. He describes this as the "first full-scale GIS war," and reminds us that many civilian Iraqi people lost their lives as the result of this military use of geospatial technologies. History is repeating itself. An Associated Press story from Baghdad on January 9, 2005, reported that "a U.S. military statement said that an F-16 jet dropped a 500-pound GPS-guided bomb on a house that was meant to be searched. . . . The house was not the intended target for the airstrike. The intended target was another location nearby" (Associated Press, 2005). The U.S. military said five people died in this tragedy, while "the man who owned the house said that the bomb killed 14 people, and an Associated Press photographer said seven of them were children" (Associated Press, 2005). Dobson (1993) encourages governments to do more to promote the peaceful use of GIS and turn their attention to cultural and social research as well as development.

Obligations to Individuals in Society

Another critical issue with respect to the obligations of GIS professionals to society is data privacy. Obermeyer (1998: 225) suggests that "in the information age, our privacy hangs on a single thread: that all the information about us that public and private organizations have gathered is housed in a large number of separate agencies that interact on a strictly limited basis. [The] ability to match such disparate databases has the potential to severely erode our privacy."

If data matching threatens individual privacy, then data mismatching is a cause for even greater concern. This is not a hypothetical situation. In the spring of 2004, Brandon Mayfield, an American Muslim lawyer, and former Army lieutenant in Portland, Oregon, was imprisoned for 2 weeks after the FBI mistakenly linked his "fingerprints to one found near the scene of a terrorist bombing in Spain. . . . Court documents released Monday suggested that the mistaken arrest first sprang from an error by the FBI's supercomputer for matching fingerprints and then was compounded by the FBI's own analysts" (MSNBC, 2004). With heightened concerns about terrorism in the United States, and the accompanying reorganization of intelligence agencies, data mismatches of this type are likely to increase, rather than decrease.

Obligations to Students

The final relationship in which ethics come into play for GIS professionals is that between GIS instructors and their students. The most obvious mandate in this relationship is that the GIS instructor do his or her best to transmit "best practice" to his or her students. This recalls the most basic of ethical issues, competence. But there is far more to the instructor–student relationship than helping students learn how to "do" GIS competently. The instructor must actively work to instill an ethical foundation in his or her students (Obermeyer, 1998).

This will not be an easy task, however, since GIS education comes in a variety of forms. GIS practitioners may receive their GIS education from an institution of higher learning as students of leading practitioners in the field who may trace their intellectual lineage to the GIS founders. Or the GIS practitioner may learn the ropes through the use of a tutorial that includes a limited-life-span copy of a particular software (e.g., "Getting to Know ArcView 9.1"). Students may take GIS courses online through universities (such as Penn State's program) or from GIS vendors. (The Environmental Systems Research Institute has an extensive set of course offerings.) There is much variation in all of these training and educational programs. It is likely, although not guaranteed, that the GIS student enrolled in an institution of higher learning will have some exposure to the ethical issues related to GIS. And Penn State is currently developing an online GIS ethics course. ESRI has been working with the GIS Certification Institute on the GIS Certificate. But there are many GIS programs that do nothing to introduce ethics.

Some authors of introductory GIS texts have adopted the practice of including a chapter on GIS ethics. While this is a good start, these chapters are also often placed on the final pages of the book. Certainly including ethics discussions is a positive development, but the end-of-course "timing comes long after students' initial excitement about learning the technology is likely to have been replaced by the need to meet end-of-semester deadlines. As a result, the cautions may fall on less-than-eager ears" (Obermeyer, 1998: 227).

Tobler's First Law of Geography states that "everything is related to everything else; but near things are more closely related than distant things" (Tobler, 1970). Embedded in this straightforward sentence is the basis for all spatial analysis and cartographic information, both of which are integral to GIS. The idea that spatial relationships exist at all is a profound insight, especially to someone who may be taking GIS training outside of the context of an integrated program in one of the spatial sciences. By pointing out that there is a correlation between distance and the strength of relationships, Tobler provides a springboard for discussion of spatial analysis and representation, including the idea of discrete and continuous attributes.

Tobler offers another useful insight into the dynamic nature of GIS as a technology when he points out that sound theory has a half-life of about 20 years, while IT has a half-life of about 5 years (probably less nowadays). This insight serves as a caveat to students who may throw themselves headlong into learning a constantly changing technology without adequately considering the nature of the tasks that the technology performs, or the models embedded in the system to perform those tasks. Students must be aware of what they are doing in a precise geographic sense, rather than merely in a software-driven command environment situation.

Along with an understanding of geographic analysis, students must also understand basic cartographic principles. Making maps using GIS is deceptively easy, which may lead students to underestimate the power of maps to inform, misinform, or disinform the map reader. This being the case, the mapmaker must take responsibility for his or her cartographic creations. Imhof (1963: 15) has raised significant concerns about the use—and potential misuse—of "technical aids" in the hands of "wannabe" (our words, not Imhof's) cartographers, and warned that "the cartographic problems concerned with the graphic composition and design of the map have increased in complexity, in spite of the new technical aids."

Obermeyer (1998) notes that GIS instructors are responsible for providing our students with the skills they need to become competent in the design, development, and use of GIS. GIS instructors must also keep in mind that their students will someday work as GIS professionals, and that they must also "consider the impact of [their] work" (Kaplan, 1986, as reported by Craig, 2004). They must understand that their actions may affect their coworkers, their professional colleagues, their clients, and society at large.

GISCI Code of Ethics

As noted, the development of the GIS Certification Institute's nationwide certification program has made possible the development and dissemination of the GIS Code of Ethics. This code was largely the work of Will Craig, and was instituted in 2005. The GISCI Code of Ethics is relatively short, and because of this, we have made the decision to include it in this chapter in its entirety, below.[1] As you will see, it does not include any reference to the professional–student relationship.

A GIS Code of Ethics

This Code of Ethics is intended to provide guidelines for GIS (geographic information system) professionals. It should help professionals make appropriate and ethical choices. It should provide a basis for evaluating their work from an ethical point of view. By heeding this code, GIS professionals will help to preserve and enhance public trust in the discipline.

This code is based on the ethical principle of always treating others with respect and never merely as means to an end: i.e., *deontology*. It requires us to consider the impact of our actions on other persons and to modify our actions to reflect the respect and concern we have for them. It emphasizes our obligations to other persons, to our colleagues and the profession, to our employers, and to society as a whole. Those obligations provide the organizing structure for these guidelines.

The text of this code draws on the work of many professional societies. It is not surprising that many codes of ethics have a similar structure and provide similar guidelines to their professionals, because they are based on a similar concept of morality. A few of the guidelines that are unique to the GIS profession include the encouragement to make data and findings widely available, to document data and products, to be actively involved in data retention and security, to show respect for copyright and other intellectual property rights, and to display concern for the sensitive data about individuals discovered through geospatial or database manipulations. Longer statements expand on or provide examples for the GIS profession.

A positive tone is taken throughout the text of this code. GIS professionals commit themselves to ethical behavior rather than merely seeking to avoid specific acts. The problems with listing acts to be avoided are: 1) there are usually reasonable exceptions to any avoidance rule and 2) there is implicit approval of any

---

[1]From GIS Certification Institute (2005). Copyright 2005 by the GIS Certification Institute. Reprinted by permission.

act not on the list. Instead, this code provides a list of many positive actions. These explicit actions illustrate respect for others and help strengthen both an understanding of this ethos and a commitment to it.

This code is not expected to provide guidelines for all situations. Ambiguities will occur and personal judgment will be required. Sometimes a GIS professional becomes stuck in a dilemma where two right actions are in conflict with each other or any course of action violates some aspect of this code. Help might come from talking with colleagues or reading relevant works such as those listed in the bibliography. Ultimately, a professional must reflect carefully on such situations before making the tough decision. Contemplating the values and goals of alternative ethical paradigms may be useful in reaching a decision:

- View persons who exemplify morality as your own guide (Virtue Ethics)
- Attempt to maximize the happiness of everyone affected (Utilitarianism)
- Only follow maxims of conduct that everyone else could adopt (Kantianism)
- Always treat other persons as ends, never merely as means (Deontology)

I. Obligations to Society

The GIS professional recognizes the impact of his or her work on society as a whole, on subgroups of society including geographic or demographic minorities, on future generations, and inclusive of social, economic, environmental, or technical fields of endeavor. Obligations to society shall be paramount when there is conflict with other obligations. Therefore, the GIS professional will:

1. Do the Best Work Possible
   - Be objective, use due care, and make full use of education and skills.
   - Practice integrity and not be unduly swayed by the demands of others.
   - Provide full, clear, and accurate information.
   - Be aware of consequences, good and bad.
   - Strive to do what is right, not just what is legal.
2. Contribute to the Community to the Extent Possible, Feasible, and Advisable
   - Make data and findings widely available.
   - Strive for broad citizen involvement in problem definition, data identification, analysis, and decision-making.
   - Donate services to the community.
3. Speak Out About Issues
   - Call attention to emerging public issues and identify appropriate responses based on personal expertise.
   - Call attention to the unprofessional work of others. First take concerns

to those persons; if satisfaction is not gained and the problems warrant, then additional people and organizations should be notified.

- Admit when a mistake has been made and make corrections where possible.

II. Obligations to Employers and Funders

The GIS professional recognizes that he or she has been hired to deliver needed products and services. The employer (or funder) expects quality work and professional conduct. Therefore the GIS professional will:

1. Deliver Quality Work
   - Be qualified for the tasks accepted.
   - Keep current in the field through readings and professional development.
   - Identify risks and the potential means to reduce them.
   - Define alternative strategies to reach employer/funder goals, if possible, and the implications of each.
   - Document work so that others can use it. This includes metadata and program documentation.
2. Have a Professional Relationship
   - Hold information confidential unless authorized to release it.
   - Avoid all conflicts of interest with clients and employers if possible, but when they are unavoidable, disclose that conflict.
   - Avoid soliciting, accepting, or offering any gratuity or inappropriate benefit connected to a potential or existing business or working relationship.
   - Accept work reviews as a means to improve performance.
   - Honor contracts and assigned responsibilities.
   - Accept decisions of employers and clients, unless they are illegal or unethical.
   - Help develop security, backup, retention, recovery, and disposal rules.
   - Acknowledge and accept rules about the personal use of employer resources. This includes computers, data, telecommunication equipment, and other resources.
   - Strive to resolve differences.
3. Be Honest in Representations
   - State professional qualifications truthfully.
   - Make honest proposals that allow the work to be completed for the resources requested.
   - Deliver an hour's work for an hour's pay.
   - Describe products and services fully.
   - Be forthcoming about any limitations of data, software, assumptions, models, methods, and analysis.

III. Obligations to Colleagues and the Profession

The GIS professional recognizes the value of being part of a community of other professionals. Together, we support each other and add to the stature of the field. Therefore, the GIS professional will:

1. Respect the Work of Others.
   - Cite the work of others whenever possible and appropriate.
   - Honor the intellectual property rights of others. This includes their rights in software and data.
   - Accept and provide fair critical comments on professional work.
   - Recognize the limitations of one's own knowledge and skills and recognize and use the skills of other professionals as needed. This includes both those in other disciplines and GIS professionals with deeper skills in critical sub-areas of the field.
   - Work respectfully and capably with others in GIS and other disciplines.
   - Respect existing working relationships between others, including employer/employee and contractor/client relationships.
   - Deal honestly and fairly with prospective employees, contractors, and vendors.
2. Contribute to the Discipline to the Extent Possible
   - Publish results so others can learn about them.
   - Volunteer time to professional educational and organizational efforts: local, national, or global.
   - Support individual colleagues in their professional development. Special attention should be given to underrepresented groups whose diverse backgrounds will add to the strength of the profession.

IV. Obligations to Individuals in Society

The GIS professional recognizes the impact of his or her work on individual people and will strive to avoid harm to them. Therefore, the GIS professional will:

1. Respect Privacy
   - Protect individual privacy, especially about sensitive information.
   - Be especially careful with new information discovered about an individual through GIS-based manipulations (such as geocoding) or the combination of two or more databases.
2. Respect Individuals
   - Encourage individual autonomy. For example, allow individuals to withhold consent from being added to a database, correct information about themselves in a database, and remove themselves from a database.
   - Avoid undue intrusions into the lives of individuals.

- Be truthful when disclosing information about an individual.
- Treat all individuals equally, without regard to race, gender, or other personal characteristic not related to the task at hand.

Further Steps to Foster Ethics among GIS Professionals

The adoption of the GIS Code of Ethics by the GIS Certification Institute (GISCI) is an important step forward. Of course, the GISCI Code of Ethics is most important to GIS professionals who choose to become certified through GISCI. However, anyone who works in GIS has access to this code of ethics. Still, words on a page (or in this case, on a website) are not enough to foster ethics among GIS professionals.

The GIS Certification is undertaking additional affirmative efforts to promote ethical behavior. The Ethics Committee of the GIS Certification Institute is actively working toward developing a workbook of GIS ethics based on real-world examples. A model for developing this workbook has been suggested. This model calls for the establishment of a process by which people could present real-world GIS ethical dilemmas to a small group of people associated with GISCI. This small group would "sanitize" the cases brought to its attention in order to remove identifying names, places, dates, and other information. The sanitized case would then be posted on a website where GIS professionals could comment and provide their insights. Eventually, these cases would be organized into a more cohesive workbook to help GIS professionals negotiate the ethical rapids of their work.

As already noted, Penn State University is in the process of developing an online GIS ethics course. This course will bring together the insights of a several GIS ethicists to design reading materials, discussion questions, and other activities that would enhance the understanding of the ethics of GIS among professionals in the field. Others, including Francis Harvey (University of Minnesota) have already put into place GIS ethics courses. GISCI itself is also considering developing short courses and workshops that could be held in conjunction with GIS and other related conferences.

The GIS Certification Institute is also considering what, if anything, it must do in cases where individuals whom it has certified engage in unethical behavior. Legal concerns related to possible restraint of trade make revocation of certification unlikely. Other options come to mind, such as requiring offending GISP professionals to take a course on ethics, or otherwise to demonstrate that they understand the wrongness of their actions, and are willing to make amends.

Conclusions

It has been nearly 20 years since Will Craig took to heart Marshall Kaplan's words, "Consider the impact of your work." We are now seeing important early steps in the evolution of GIS ethics, with the creation and dissemination of the GIS Code of Ethics. Turning this code into action, and bringing it to GIS practitioners more widely will take a great deal of work.

Books, coursework, training, and other avenues will be needed to promote GIS ethics more broadly. The champions and the motivation exist to make this happen. While we do not know exactly what will be the outcome, we know that GIS ethics will become increasingly important in the coming years.

Chapter 15

Envisioning a Future

In the nearly 15 years since the first edition of this book was published, the field of GIS has changed dramatically. Much of this change is attributable simply to the advancement of the technology related to the acquisition and use of geographic information, as systems become more complex, dynamic, and technologically sophisticated. Thus, though it would be valuable to recognize the natural changes that time and technology have rendered on the field, it would nevertheless be a case of recognizing the obvious. Of far more interest to us, however, have been the concomitant changes that have occurred over the past 15 years in terms of *consequences* of GIS advancement, both intended and unintended. Technological change is always a precursor to a myriad of sociological, behavioral, and political effects. In this manner, GIS is no different than any other field. Technology begets its own implications.

Because the authors come at the field of GIS from different backgrounds, we tend to naturally recognize and focus on those issues in managing GIS that are of most immediate importance to us. Having a strong background in behavioral theory and organizational dynamics leads one of us (Jeffrey Pinto) to focus on the realm of the organizational implications of GIS technology. This focus includes its effects on power and political realities; the best means for implementing these systems; the manner in which organizations employ GIS as a driver of and response to larger, strategic goals; the reasons to decide whether or not to share geographic information; and so forth. In short, one important consequence of the

313

rapid development of GIS has been the organizational and behavioral implications that this technology spawns.

Another distinct, but equally relevant, stream of interest and implication of GIS development and proliferation is the manner in which GIS has influenced public policy, legal rights and responsibilities, and ethical considerations. While we have discussed each of these issues in detail in the preceding chapters, it is worthwhile to remind ourselves that, as with any advance of new technologies, the law of unintended consequences come very much into play. That is, as the wave of GIS continues to rise, as more and more uses are discovered for geographic information, we uncover new and fertile grounds for debate. Does the capacity to do something new automatically trigger the expectation to follow where the technology leads us? In short, new technologies will always demand new thinking.

A third stream of "effect" from advances in GIS technology lies in the professionalization of the field, as individuals develop careers in city planning, emergency response, land development and reclamation, and so forth. GIS have had a tremendous impact on the manner in which these professionals do their jobs and the new technical requirements necessary to maintain standards of practice. Thus, while we have explored the "higher" order implications for managing GIS in this book (e.g., organizational, behavioral, public policy, legal), we cannot ignore how individuals are impacted by GIS within their work settings, including the manner in which GIS advances are demanding ever greater technical sophistication and advanced learning. Chapter 12, on ensuring the qualifications of GIS professionals, speaks directly to this new-felt requirement.

A distinct challenge of management lies in the need to simultaneously focus on the minute and the broad, in effect, to understand the "trees" while also having the ability to adopt a 10,000-foot view. Geographic information technology is a case in point for this need. Successful and effective GIS managers must be conversant with the power of GIS, while operating within the strict (some would say "limiting") confines of the home organization. Whether working in an urban planning office, a charity, an emergency rapid-response organization, or some other organization for which geographic information is pertinent and necessary, GIS managers must first understand the technology they are supporting. Therefore, at a minimum, it seems to us that effective GIS managers recognize the power of the tool at hand. However, this technical mastery is by no means the guarantee of success. Without an equally clear-eyed knowledge of organizational dynamics and human behavior, the most technically savvy person in the world will fall short of fulfilling his or her career aspirations.

The "trick," in our opinion, lies in recognizing a salient truth of organizational life in any setting: the need to achieve a balance between the technical and the individual, between the science and the facts of human interaction. Thus, the most technically qualified person can routinely fail if he or she does not understand the political framework of any organization, the basic implications of motivation and leadership, and the manner in which any meaningful organizational change is ever likely to be accepted and diffused. These, in short, are not technical questions. They are, and will always remain, people challenges first.

Managing Geographic Information Systems was originally conceived and written when the field was much younger and rapidly exploring the means to achieve various goals and ends. We welcomed the advance of geographic information technology because it offered (and still offers) so many opportunities and challenges that it seems as though every new breakthrough opens even more vistas for future advances. These advances have not come without their own policy implications, however, and that has been one of the original drivers of this book (and still motivates us to this day).

Looking down the road, several important trends seem likely to continue to push GIS down the road of technological, organizational, professional, historical, political, and societal change. Technological trends include both the continuing development of hardware and software for the capture, storage, analysis, and presentation of geographic data. Organizational trends include greater empowerment and freedom of information, which the governments of some countries (notably China) resist and actively thwart through technological modifications of their own (with the help of multinational conglomerates such as Cisco, Microsoft, Google, and Yahoo). Certainly, historical events will continue to play a role in the use of GIS and cognate technologies; we have already seen the terrorist acts of September 11, 2001, hastening the use of geographic tracking devices. Finally, all of these elements together will lead to additional political and societal change. Indeed, all five of these elements are closely intertwined and developments in each element profoundly affects the others.

In the coming years, GIS will continue to become more powerful, easier to use, and more portable. GIS software will continue to become more accessible and less expensive, spurred in part by the continuing evolution of open-source GIS software. The Internet is an integral part of this development. Many legitimate websites offer open-source GIS software to potential users. Several such software packages, such as GEODA, are already available for free downloading. Other free online sources, such as the "GIS Cookbook," are also available to help the novice use GIS. Several GIS scholars, notably Harlan Onsrud at the University of Maine, make Powerpoint presentations of their GIS courses available online too. The increasing availability of such low-cost (and in some cases no-cost), high-value GIS resources widens the opportunities for the use of GIS as a tool for empowerment, making it more and more feasible for previously marginalized groups to make use of GIS technology to further their own policy goals.

Indeed, the community mapping or public participation movement is strong and growing all the time. Community groups, environmental organizations, and other special interest groups are harnessing GIS to help them make their case to their neighbors, their political rivals, and their elected officials. Thanks to the willingness of many institutions—especially government agencies—to share their data sets, community mapping groups have basemaps into which they can add their own data and build their own projects, using their local knowledge to add value to the data sets. Or, if they find inaccuracies in these official basemaps, community mapping groups can attempt to correct them. Should there be a third edition of *Managing Geographic Information Systems*, greater exploration of the successes and failures of community mapping and GIS on the Internet would be a valuable addition to our efforts.

Existing cognate technologies, such as GPS, radio frequency identification (RFID) devices, and closed-circuit television will become more common in the years ahead. Already they are readily available through traditional bricks-and-mortar retail establishments, but also online. These technologies bring with them increased capacities for geographic surveillance, both open and aboveboard and clandestine. The triangulation associated with traditional cellphones provides a general location for users, while the newest models embed GPS technology, which provides a more specific location for the caller.

This technological change does have its advantages. For example, the newest model cellphones allow Emergency-911 first responders to know within several meters where you are when you slide off an icy road, enabling them to get to you more quickly. But there is a distinctly sinister downside to the improvement in locational precision. GPS technology can be used to track individuals who may not wish to be found. While an RFID on a difficult-to-remove bracelet alerts hospital staff when an unauthorized individual moves the baby, how will that child feel 17 years later when Mom and Dad place a data-recording device capable of reporting speeding or hard-braking incidents on the family car that he or she drives around town? This parental use would mostly be perceived as still within reason, certainly by other parents of young drivers (especially if they are footing the bill for car insurance), even if the young driver dislikes the invasion of privacy. A more clear-cut invasion of privacy occurs when a stalker places the device on the vehicle of the target of his or her obsession. It will be necessary to keep abreast of these developments, along with the legal decisions and legislation that accompany them, as a GIS management issue.

Professional change within the GIS community is already a reality. Although far from ubiquitous, certification of GIS professionals is becoming more common. We have discussed the growing emphasis on ethics within the GIS community in a previous chapter, but there are other intriguing trends. For example, in Australia, the Spatial Sciences Institute (SSI) has instituted a GIS certification program based, in part, on the program implemented by the GIS Certification Institute in the United States. In spite of these first steps in GIS certification, there remain many highly qualified and ethical GIS practitioners who question the value of GIS certification; these individuals play a valuable role in helping to improve certification as they raise questions about it.

Another trend that is encouraging the professionalization of the GIS community is the Model Curricula Body of Knowledge developed under the auspices of the University Consortium for Geographic Information Science (UCGIS). This initiative has been in process for well over a decade, beginning with the "Core Curriculum" of the National Center for Geographic Information and Analysis in the early 1990s. This long-term project has relied on the expertise of scholars, educators, and other members of the GIS community to identify key issues in GIS, very broadly construed. Altogether, these and other similar initiatives contribute to a GIS community with greater consensus on its knowledge base and on what it takes to become a qualified and responsible GIS professional.

History-making events also play a role in the evolution of GIS and the manner in which organizations use them. Certainly the terrorist attacks of September 11, 2001, in the United States and subsequent attacks on transit systems in London and Madrid have spurred some policymakers to implement additional technologies for tracking the geographic location of their citizens and residents. More, not less, tracking is likely to occur in the future.

It is not only governments whose GIS directions are influenced by these historical events. Penn State University is preparing to offer a postbaccalaureate certificate in geospatial intelligence (see *ist.psu.edu/ prospectivestudents/undergraduate/sra/*), and several other institutions of higher education across the United States are in the development stage with similar programs. Certainly, this raises significant concerns regarding the ethical and privacy considerations associated with GIS implementation. We should add that educators and administrators within Penn State's GIS program have been playing a leading role in ethics education for GIS professionals, but not all proposed programs have such an orientation. This is another trend that warrants continued monitoring.

Even before 9/11, GIS scholars, educators, and implementers have been responding to our changing world. In particular, GIS has long been used as a tool in environmental analysis and remediation efforts. Whatever history brings, GIS will be there to help us understand our changing world and to attempt to mitigate unfortunate outcomes. Sometimes its presence will be thoughtful and valuable. Other times, its presence will be a knee-jerk response that may trigger unforeseen negative consequences.

Finally, political and societal changes also play a role in the continuing evolution of GIS technology, sometimes in contradictory ways. On the one hand, we recognize the growing use of GIS as a tool to enhance what Dobson (2003) calls "geoslavery." On the other hand, we acknowledge the valuable role that GIS has played in empowering previously marginalized groups, and enlarging discussions surrounding policy, especially on the local and regional levels. In the future, we will likely see continuing development of the technology both as a tool of control and a tool of empowerment.

Interpretation of laws and passage of new legislation at a variety of levels from the city and county to the state (country) and even within international rule-making bodies will help define the limits of GIS. At the same time, entrepreneurs may see a market for and develop countervailing technologies to limit the effects of geographic surveillance devices. As people without vast expertise in GIS, but with a solid foundation of local knowledge, become more familiar with geographical information technologies and resources, they may help to influence laws, legislation, and policy.

And then the cycle begins again, as the technology developers take steps to improve the technology for an ever widening user base. Today, geographic information technology users include not just organizations and people on a mission; it increasingly includes people and organizations using the technology in a recreational context. Popular websites like Google Earth and Mapquest introduce everyday Internet users to the fun and practical value of spatial data. For the more advanced and adventurous, geocaching websites provide another recreational application of GPS, a cognate technology. These casual points of entry to GIS and related geographic information technology have tremendous potential to draw individuals to the field who previously would not even have known that GIS exists. Thus, there is some potential for a greater dynamism and democratization of GIS and its implementation in the future and for more creative applications down the road. We will continue to keep our eyes on these and other evolving trends.

We firmly believe, with the publication of this second edition of our book, that the "heyday" of geographic information still lies in the future. As we progress optimistically into that future, we need to carry with us an understanding of the means to gain the most we can from GIS technologies. Organizational and technological innovations do not generally trigger their own benefits; their capabilities must be carefully managed. It was with this goal in mind that this book was first written and has been updated. It is our hope that readers will find something useful and applicable in each chapter as they work to make successful GIS a reality in their organizations.

References

Aitken, S. C., & Michel, S. M. (1995). Who contrives the "real" in GIS?: Geographic information, planning, and critical theory. *Cartography and Geographic Information Systems, 22*(1), 17–29.

Alfelor, R. M. (1995). GIS and the integrated highway information system. In H. J. Onsrud & G. Rushton (Eds.), *Sharing geographic information* (pp. 397–412). New Brunswick, NJ: Center for Urban Policy Research.

Al-Kodmany, K. (2002). GIS and the artist: Shaping the image of a neighbourhood through participatory environmental design. In W. J. Craig, T. M. Harris, & D. Weiner (Eds.), *Community participation and geographic information systems* (pp. 320–329). New York: Taylor & Francis.

Allison, G. T. (1971). *The essence of decision.* Boston: Little, Brown.

Anderson, J. C., & Narus, J. A. (1986). Toward a better understanding of distribution channel working relationships. In K. Backhaus & D. Wilson (Eds.), *Industrial marketing: A German-American perspective* (pp. 320–336). Berlin: Springer-Verlag.

Anselin, L. (2002, April). *Use of GIS for spatial analysis.* Presentation before the National Research Council of the National Academy of Sciences Committee on Use of GIS by the U.S. Department of Housing and Urban Analysis, Washington, DC.

Antenucci, J. C., Brown, K., Croswell, P. L., & Kevany, M. (1991). *Geographic information systems: A guide to the technology.* New York: Van Nostrand Reinhold.

Archer, H., & Croswell, P. L. (1989). Public access to geographic information systems: An emerging legal issue. *Photogrammetric Engineering and Remote Sensing, 55,* 1575–1581.

Argote, L. (1982). Input uncertainty and organizational coordination in hospital emergency units. *Administrative Science Quarterly, 27,* 420–434.

Arnoff, E. L. (1971). Successful models I have known. *Decision Sciences, 2,* 141–148.

Aronoff, S. (1989). *Geographic information systems: A management perspective.* Ottawa, Ontario, Canada: WDL Publications.

Associated Press. (2005, January 9). Iraq in transition: U.S. bombs wrong house, killing at least five. *Hoosier Times,* p. A3.

Astley, W. G., Axelsson, R., Butler, R., Hickson, D., & Wilson, D. (1982). Complexity and cleavage: Dual explanations of strategic decision-making. *Journal of Management Studies, 19*(4), 357–365.

Azad, B. (1997). High and low roads to GIS development. Qatar Conference on GIS. Available at *www.gisqatar.org.qa/conf97/links/j4.html.*

Azad, B. (1998). *Management of enterprise-wide GIS implementation: Lessons from exploration of five case studies.* Cambridge, MA: MIT Press.

Azad, B., & Wiggins, L. L. (1995). Dynamics of inter-organizational geographic data sharing: A conceptual framework for research. In H. J. Onsrud & G. Rushton (Eds.), *Sharing geographic information* (pp. 22–43). New Brunswick, NJ: Center for Urban Policy Research.

Bacharach, S. B., & Lawler, E. (1980). *Power and politics in organizations.* San Francisco: Jossey-Bass.

Bean, A. S., Neal, R. D., Radnor, M., & Tansik, D. A. (1975). Structural and behavioral correlates of implementation in U.S. business organizations. In R. L. Schultz & D. P. Slevin (Eds.), *Implementing operations research and management science* (pp. 77–132). New York: Elsevier.

Beard, M. K., & Buttenfield, B. P. (1999). Detecting and evaluating errors by graphical methods. In P. A. Longley, M. F. Goodchild, D. J. Maguire, & D. W. Rhind (Eds.), *Geographical information systems: Vol. 1. Principles and technical issues* (2nd ed., pp. 219–233). New York: Wiley.

Benbasat, I., Goldstein, D. K., & Mead, M. (1987). The case research strategy in studies of information systems. *MIS Quarterly, 11*(3), 369–386.

Bennett, R. J. (1980). *The geography of public finance.* London: Methuen.

Benveniste, G. (1989). *Mastering the politics of planning: Crafting credible plans and policies that make a difference.* San Francisco: Jossey-Bass.

Berlant, J. L. (1975). *Profession and monopoly: A study of medicine in the United States and Great Britain.* Berkeley: University of California Press.

Bernknopf, R. L., Brookshire, D. S., Soller, D. R., McKee, M. J., Sutter, J. F., Matti, J. C., et al. (1993). *Societal value of geological maps.* Washington, DC: U.S. Geological Survey.

Berry, J. (1994, September 5). Potent new tool for selling: Database. *Business Week,* pp. 56–62.

Bie, S. (1984). Organizational needs for technological advancement. *Cartographica, 21,* 44–50.

Blau, P. M. (1970). Decentralization in bureaucracies. In M. Zald (Ed.), *Power in organizations.* Nashville, TN: Vanderbilt University.

Bond, C. (2002). The Cherokee Nation and tribal uses of GIS. In W. J. Craig, T. M. Harris, & D. Weiner (Eds.), *Community participation and geographic information systems* (pp. 283–294). New York: Taylor & Francis.

Bonoma, T. V. (1985). Case research in marketing: Opportunities, problems, and a process. *Journal of Marketing Research, 22,* 199–208.

Bozeman, B. (1987). *All organizations are public: Bridging public and private organizational theories.* San Francisco: Jossey-Bass.

Bozeman, B., & Bretschneider, S. (1986). Public management information systems: Theory and prescription. *Public Administration Review, 46* [Special issue], 475–487.

Bozeman, B., & Loveless, S. (1987). Sector context and performance: A comparison of industrial and government research units. *Administration and Society, 19,* 197–235.

Bretschneider, S. (1990). Management information systems in public and private organizations: An empirical test. *Public Administration Review, 50,* 536–545.

Brill, E. D., Flach, J. M., Hopkins, L. D., & Ranjithan, S. (1990). MGA: A decision support system for complex, incompletely defined problems. *IEEE Transactions on Systems, Man and Cybernetics, 20,* 745–757.

Brown, M. M., & Brudney, J. L. (1993). *Modes of geographical information system adoption in public organizations: Examining the effects of different implementation structures.* Paper presented at the annual meeting of the American Society for Public Administration.

Brown, M. M., Brudney, J. L., & O'Toole, L. J., Jr. (1998). Implementing information technology in government: An empirical assessment of the role of local partnerships. *Journal of Public Administration Research and Theory, 8*(3), 449–525.

Brubaker, R. (1984). *The limits of rationality: An essay on the social and moral thought of Max Weber.* London: Allen & Unwin.

Budic, Z. (1993). GIS use among southeastern local governments. *URISA Journal, 5*(1), 4–17.

Buehler, K., & McKee, L. (1996). *The Open GIS guide.* Wayland, MA: The Open GIS Consortium.

Buntz, C. G., & Radin, B. A. (1983). Managing intergovernmental conflict: The case of human services. *Public Administration Review, 43,* 403–412.

Burns, T., & Stalker, G. M. (1961). *The management of innovation.* London: Tavistock.

Burroughs, P. A., & McDonnell, R. A. (1998). *Principles of geographical information systems.* New York: Oxford University Press.

Byrnes, N. (2005, October 31). Leader of the packs: Marlboro still smokin' at 50, thanks to buzz marketing. *Business Week.* Available at *www.businessweek.com/print/magazine/content/05_44/b3957107.htm?chan+gl*

Calkins, H. W., & Weatherbe, R. (1995). Taxonomy of spatial data sharing. In H. J.

Onsrud & G. Rushton (Eds.), *Sharing geographic information* (pp. 65–75). New Brunswick, NJ: Center for Urban Policy Research.

Camillus, J. C. (1986). *Strategic planning and management control*. Lexington, MA: Lexington Books.

Campbell, H. J. (1991). *Organizational issues and the utilization of geographic information systems* (Regional Research Laboratory Initiative Discussion Paper No. 9). Sheffield, UK: Economic and Social Research Council, Regional Research Laboratories, University of Sheffield.

Campbell, H. J. (1993). GIS implementation in British local government. In I. Masser & H. J. Onsrud (Eds.), *Diffusion and use of geographic information technologies* (pp. 117–146). Dordrecht, The Netherlands: Kluwer.

Campbell, H. J., & Masser, I. (1991). The impact of GIS on local government in Great Britain. In *Proceedings of the Association for Geographic Information Conference*. London: AGI.

Casey, L., & Pederson, T. (2002). Mapping Philadelphia's neighbourhoods. In W. J. Craig, T. M. Harris, & D. Weiner (Eds.), *Community participation and geographic information systems* (pp. 65–76). New York: Taylor & Francis.

Cash, J. I., Jr., Eccles, R. G., Nohria, N., & Nolan, R. L. (1994). IT between organizations: Interorganizational systems. In J. I. Cash, Jr., R. G. Eccles, N. Nohria, & R. L. Nolan (Eds.), *Building the information-age organization: Structure, control, and information technologies* (3rd ed., pp. 338–396). Burr Ridge, IL: Irwin.

Catsambas, T. (1982). Substitutability, separability, and the distributional implications of public goods. *Public Finance Quarterly, 10*, 333–353.

Cayer, N. J., & Weschler, L. F. (1988). *Public administration: Social change and adaptive management*. New York: St. Martin's Press.

Cho, G. (1998). *Geographic information systems and the law: Mapping the legal frontiers*. New York: Wiley.

Chrisman, N. R. (1987). Design of geographic information systems based on social and cultural goals. *Photogrammetric Engineering and Remote Sensing, 53*, 1367–1370.

Christie, R., & Geis, F. (1970). *Studies in Machiavellianism*. New York: Academic Press.

Churchman, C. W., & Schainblatt, H. A. (1965). The researcher and the manager: A dialectic of implementation. *Management Science, 11*, 1369–1387.

Citera, M., McNeese, M. D., Brown, C. E., Selvaraj, J. A., Zaff, B., & Whitaker, R. D. (1995). Fitting information systems to collaborating design teams. *Journal of the American Society for Information Science, 46*(7), 551–559.

Clark, G. L. (1981). Law, the state, and the spatial integration of the United States. *Environment and Planning A, 13*(10), 1197–1232.

Clarke, K. C. (2002). *Getting started with GIS* (4th ed.). Upper Saddle River, NJ: Prentice Hall.

Clemons, E., & Kenz, M. (1988). Competition and cooperation in information systems innovation. *Information and Management, 14*(1), 25–35.

Clemons, E., & Row, M. (1992). Information technology and industrial cooperation: The changing economics of coordination and ownership. *Journal of Management Information Systems, 9*(2), 9–28.

Cleveland, H. (1985). The twilight of hierarchy: Speculations on the global information society. *Public Administration Review, 45*(2), 185–195.

Cook, T. D., & Campbell, D. T. (1979). *Quasi-experimentation: Design and analysis for field settings.* Boston: Houghton Mifflin.

Corey, M. (1998). Digital data, copyright, commercialization, and related issues: A Canadian perspective. In D. R. F. Taylor (Ed.), *Policy issues in modern cartography* (pp. 29–46). New York: Pergamon Press.

Craig, W. J. (1993). A GIS code of ethics: What can we learn about from other organizations? *Journal of the Urban and Regional Information Association, 5*(2), 13–16.

Crampton, J. W. (1995). The ethics of GIS. *Cartography and Geographic Information Systems, 22*(1), 29–40.

Crampton, J. W. (2006). *Surveillance, security and personal dangerousness.* Paper presented on the panel *Orwell's wolf is back: Tracking kids, dogs, old people and everybody in between.* St. Louis, MO: American Association for the Advancement of Science.

Croswell, P. L. (1989). Facing reality in GIS implementation: Lessons learned and obstacles to be overcome. *Proceedings of the 27th Annual Urban and Regional Information Systems Association, 4,* 15–35.

Croswell, P. L. (1991). Obstacles to GIS implementation and guidelines to increase the opportunities for success. *Journal of the Urban and Regional Information Systems Association, 3*(1), 43–56.

Crozier, M. (1964). *The bureaucratic phenomenon.* London: Tavistock.

Curry, M. R. (1995). Rethinking rights and responsibilities in geographic information systems: Beyond the power of the image. *Cartography and Geographic Information Systems, 22*(1), 58–69.

Cyert, R. M., & March, J. G. (1963). *A behavioral theory of the firm.* Englewood Cliffs, NJ: Prentice Hall.

Dahlgren, M. W., & Gotthard, A. (1994). *Cost–benefit analysis for information technology projects.* Washington, DC: International City Management Association.

Dansby, B. (1991). Recovering GIS development costs by copyright use. *GIS World, 4*(2), 100–101.

Danziger, J. N., Dutton, W., Kling, R., & Kraemer, K. (1982). *Computers and politics: High technology in American local government.* New York: Columbia University Press.

Deacon, R., & Shapiro, P. (1975). Private preference for collective goods revealed through voting on referenda. *American Economic Review, 64,* 943–955.

De Blij, H., & Muller, P. O. (1992). *Geography: Regions and concepts* (Rev. 6th ed.). New York: Wiley.

DeLone, W. H., & McLean, E. R. (1992). Information systems success: The quest for the dependent variable. *Information Systems Research, 3,* 60–95.

Deshpande, R., & Zaltman, G. (1987). A comparison of factors affecting use of marketing information in consumer and industrial firms. *Journal of Marketing Research, 21,* 114–118.

Deutsch, M. (1958). Trust and suspicion. *Journal of Conflict Resolution, 2,* 265–279.

Dickinson, H. J., & Calkins, H. W. (1988). The economic evaluation of implementing a GIS. *International Journal of Geographical Information Systems, 2*(4), 307–327.

Dickinson, H. J., & Calkins, H. W. (1990). Comment on "Concerning the economic evaluation of implementing a GIS." *International Journal of Geographical Information Systems, 4*(2), 211–212.

Dobson, J.E. (1993). The geographic revolution: A retrospective on the age of automated geography. *Professional Geographer, 45*(4), 431–439.

Dobson, J. E. (2003, March 5). *Geoslavery: How the global positioning system will track everyone, all the time.* Available at *http://www.oilempire.us/geoslavery.html*

Doll, W. J., & Torkzadeh, G. (1988). The measurement of end user computing satisfaction. *MIS Quarterly, 12,* 259–276.

Douglas, M. (1986). *How institutions think.* Syracuse, NY: Syracuse University Press.

Downs, A. (1967a). *Inside bureaucracy.* Boston: Little, Brown.

Downs, A. (1967b). A realistic look at the final payoffs from urban data systems. *Public Administration Review, 27*(3), 204–210.

Downs, G. W., & Mohr, L. B. (1976). Conceptual issues in the study of innovations. *Administrative Science Quarterly, 21,* 700–714.

Drory, A., & Romm, T. (1990). The definition of organizational politics: A review. *Human Relations, 43*(11), 1133–1154.

Dumaine, B. (1990, May 7). Who needs a boss? *Fortune,* pp. 52–60.

Dwyer, F. R., & Oh, S. (1987). Output sector munificence effects on the internal political economy of marketing channels. *Journal of Marketing Research, 24,* 347–358.

Ehrenreich, B. (1990). *Fear of falling: The inner life of the middle class.* New York: HarperPerennial.

Ellul, J. (1964). *The technological society.* New York: Vintage Books.

Elwood, S. (2002). The impact of GIS use for neighbourhood revitalization in Minneapolis. In W. J. Craig, T. M. Harris, & D. Weiner (Eds.), *Community participation and geographic information systems* (pp. 77–88). New York: Taylor & Francis.

Engle, J. R. (1983). *Sacred sands: The struggle for community at the Indiana Dunes.* Middletown, CT: Wesleyan University Press.

Epstein, E. F. (1990). Access to information: Legal issues. *Proceedings of the XIX Congress of the International Federation of Surveyors, 3,* 92–99.

ESRI. (2004). *Getting to Know ArcView 9.1*. Redlands, CA: ESRI Press.

Evans, J. D. (1995). *A case study of infrastructures for sharing geographic information among environmental agencies*. Paper presented at the NCGIA Young Scholars Summer Institute on Geographic Information.

Evans, J. D., & Ferreira, J., Jr. (1995). Sharing spatial information in an imperfect world: Interactions between technical and organizational issues. In H. J. Onsrud & G. Rushton (Eds.), *Sharing geographic information* (pp. 448–460. New Brunswick, NJ: Center for Urban Policy Research.

Faulhaber, G. R. (1975). Cross-subsidization: Pricing in public enterprises. *American Economic Review, 65*(5), 966–977.

Fayol, H. (1929). *General and industrial management* (J. A. Conbrough, Trans.). Geneva: International Management Institute.

Federal Geographic Data Committee (FGDC). (1994). *Development of a National Digital Geospatial Data Framework*. Status Report from the Framework Working Group. Reston, VA: Author.

Fayol, H. (1949). *General and industrial administration*. London: Pitman. (Original work published 1916)

Federal Geographic Data Committee (FGDC). (2005). Geospatial standards. Available at *fgdc.gov/publications/documents/standards/geospatial_standards_ part1.html*

Ferris, G. R. (1992). Perceptions of organizational politics. *Journal of Management*. Available at *www.findarticles.com/p/articles/mi_m4256/is_n1_v18/ai_12289739/print*

Ferris, G. R., & Kacmat, K. M. (1992). Perceptions of organizational politics. *Journal of Management, 18*(1), 93–116.

Field, B. C. (1994). *Environmental economics: An introduction*. New York: McGraw-Hill.

Filley, A. C. (1975). *Interpersonal conflict resolution*. Glenview, IL: Scott Foresman.

Fischer, F., & Forester, J. (Eds.). (1993). *The argumentative turn in policy analysis and planning*. Durham, NC: Duke University Press.

Forester, J. (1989). *Planning in the face of power*. Berkeley: University of California Press.

Fotheringham, S., & Rogerson, P. (Eds.). (1995). *Spatial analysis and GIS*. New York: Taylor & Francis.

Frank, A. U. (1992). Telecommunication and GIS: Opportunities and challenges. In P. W. Newton, P. R. Zwart, & M. E. Cavill (Eds.), *Networking spatial information systems* (pp. 235–250). London: Belhaven.

Frantz, C. R., & Robey, D. (1984). An investigation of user-led system design: Rational and political perspectives. *Communications of the ACM, 27*(12), 1202–1209.

Gage, R. W. (1984). Federal regional councils: Networking organizations for policy management in the governmental system. *Public Administration Review, 44*, 134–144.

Galbraith, J. R., & Nathanson, D. A. (1978). *Strategic implementation: The role of structure and process*. Dallas, TX: Business Publications.

Galletta, D., & Lederer, A. L. (1989). Some cautions on the measurement of user information satisfaction. *Decision Sciences, 20*, 419–438.

Gandz, J., & Murray, V. V. (1980). Experiences of workplace politics. *Academy of Management Journal, 23*, 237–251.

Garreau, J. (1981). *The nine nations of North America*. Boston: Houghton Mifflin.

George, A., & McKeown, T. (1985). Case studies and theories of organizational decision making. In L. Sproull & P. Larkey (Eds.), *Information processing in organizations* (pp. 21–58). Greenwich, CT: JAI Press.

Gersmehl, P. (1985). The map, the data, and the innocent bystander: A parable for map users. *Professional Geographer, 37*(3), 329–334.

Gerth, H. H., & Mills, C. W. (1976). *From Max Weber: Essays in sociology* (H. H. Gerth and C. W. Mills, Eds. & Trans.). New York: Oxford University Press. (Original work published 1946)

Gillespie, S. R. (1991). Measuring the benefits of GIS use. *Proceedings of the ACSM-ASPRS Fall Convention*, pp. A-84–94.

Gillroy, J. M. (1992). The ethical poverty of cost–benefit methods: Autonomy, efficiency, and public policy choice. *Policy Sciences, 25*(2), 83–102.

Ginzberg, M. J. (1978). Finding an adequate measure of OR/MS effectiveness. *Interfaces, 8*(4), 59–62.

GIS Certification Institute. (2005). *A GIS code of ethics*. Available at *www.gisci.org/code_of_ethics.htm*

GIS World Staff. (1993). Ease of use, lower costs highlight industry trends. *GIS World, 6*(12), 36–47.

GIS World Staff. (1994). GIS in business moves into the mainstream. *GIS World, 7*(5), 48.

Godschalk, D. R., Bollen, S., Hekman, J., & Miles, M. (1985). *Land supply monitoring: A guide for improving public and private urban development decisions*. Boston: OGH.

Golembiewski, R. T. (1984). Organizing public work, round three: Toward a new balance between political agendas and management perspectives. In R. T. Golembiewski & A. B. Wildavsky (Eds.), *The costs of Federalism* (pp. 237–270). New Brunswick, NJ: Transaction.

Golembiewski, R. T., & Wildavsky, A. B. (Eds.). (1984). *The costs of federalism: In honor of James W. Fesler*. New Brunswick, NJ: Transaction.

Goodchild, M. F. (1988). Stepping over the line: Technological constraints and the new cartography. *American Cartographer, 15*(3), 311–319.

Goodchild, M. F. (1993). Ten years ahead: Dobson's automated geography in 1993. *Professional Geographer, 45*(4), 444–446.

Goodchild, M. F. (2005). *Geographical information science fifteen years later*. Available at *www.ucsb.edu/˜good/papers/424.pdf*

Goodchild, M. F., & Getis A. (1991). *Introduction to spatial analysis*. Notes to accompany a workshop of the same name held in conjunction with GIS/LIS '91, Atlanta, GA.

Goodchild, M. F., & Longley, P. A. (1999). The future of GIS and spatial analysis. In P. A. Longley, M. F. Goodchild, D. J. Maguire, & D. W. Rhind (Eds.), *Geographical information systems: Vol. 1. Principles and technical issues* (2nd ed., pp. 567–580). New York: Wiley.

Goodchild, M. F., & Rizzo, B. R. (1987). Performance evaluation and work-load estimation for geographic information systems. *International Journal of Geographic Information Systems, 1*(1), 67–76.

Gould, S. J. (1983). The titular bishop of Titiopolis. In S. J. Gould, *Hen's teeth and horse's toes* (pp. 69–78). New York: Norton.

Gouldner, A. W. (1954). *Patterns of industrial bureaucracy.* New York: Free Press.

Greer, A. L. (1981). Medical technology: Assessment, adoption, and utilization. *Journal of Medical Systems, 5,* 129–145.

Grimshaw, D. J. (1994). *Bringing geographical information systems into business.* Essex, UK: Longman Scientific & Technical.

Gupta, A., Raj, S. P., & Wilemon, D. (1986). A model for studying the R&D–marketing interface in the product innovation process. *Journal of Marketing, 50,* 7–17.

Guptill, S. C. (1999). Metadata and data catalogues. In P. A. Longley, M. F. Goodchild, D. J. Maguire, & D. W. Rhind (Eds.), *Geographical information systems: Vol. 1. Principles and technical issues* (2nd ed., pp. 677–692). New York: Wiley.

Gutmann, E. (2004). *Losing the new China.* San Francisco: Encounter Books.

Habermas, J. (1970). *Toward a rational society: Student protest, science, and politics* (J. J. Shapiro, Trans.). Boston: Beacon Press.

Habermas, J. (1985). *The theory of communicative action: Vol. 1. Reason and the rationalization of society* (Thomas McCarthy, Trans.). Boston: Beacon Press.

Hardy, C. (1987). The contribution of political science to organizational behavior. In J. W. Lorsch (Ed.), *Handbook of organizational behavior* (pp. 96–108). Englewood Cliffs, NJ: Prentice Hall.

Harris, L. E. (1998). Copyright issues in modern cartography. In D. R. F. Taylor (Ed.), *Policy issues in modern cartography* (pp. 71–89). New York: Pergamon Press.

Harris, T. M., & Weiner, D. (2002). Implementing a community-based GIS: Perspectives from South African fieldwork. In W. J. Craig, T. M. Harris, & D. Weiner (Eds.), *Community participation and geographic information systems* (pp. 246–258). New York: Taylor & Francis.

Harvey, A. (1970). Factors making for implementation success and failure. *Management Science, 16,* B312–B321.

Harvey, F. (1997). Improving multi-purpose GIS design: Participative design. In S. C. Hirtle & A. U. Frank (Eds.), *Spatial information theory.* Berlin: Springer-Verlag.

Heuvelink, G. B. M. (1999). Propagation of error in spatial modeling with GIS. In P. A. Longley, M. F. Goodchild, D. J. Maguire, & D. W. Rhind (Eds.), *Geographical information systems: Vol. 1. Principles and technical issues* (2nd ed., pp. 207–217). New York: Wiley.

Heywood, I., Cornelius, S., & Carver, S. (1998). *An introduction to geographical information systems.* Upper Saddle River, NJ: Prentice Hall.

Higgs, G. (1999). Sharing environmental data across organizational boundaries: Lessons from the Rural Wales Terrestrial Database Project. *Annals of Regional Science, 33,* 233–249.

Hori, H. (1975). Revealed preferences for public goods. *American Economic Review, 65*(5), 978–991.

Hult, K. M., & Walcott, C. (1990). *Governing public organizations: Politics, structures, and institutional design.* Pacific Grove, CA: Brooks/Cole.

Hunt, S., & Nevin, J. R. (1974). Power in a channel of distribution: Sources and consequences. *Journal of Marketing Research, 11,* 186–193.

Hutchinson, S. E., & Sawyer, S. C. (1992). *Computers: The user perspective.* New York: Irwin.

Huxhold, W. E. (1991). *An introduction to urban geographic information systems.* New York: Oxford University Press.

Huxhold, W. E., & Levinsohn, A. G. (1995). *Managing information systems projects.* New York: Oxford University Press.

Igbaria, M., & Nachman, S. A. (1990). Correlates of user satisfaction with end user computing: An exploratory study. *Information and Management, 19,* 73–82.

Imhof, E. (1963). Tasks and methods of theoretical cartography. *International Yearbook of Cartography, 3,* 13–23.

Indiana Geological Survey. (2003, August). *Don't duck metadata!* Paper presented at the IGS Metadata Workshop, Indiana University, Bloomington.

Ives, B., & Olson, M. H. (1983). User involvement and MIS success: A review of research. *Management Science, 30,* 586–603.

Ives, B., Olson, M. H., & Baroudi, J. J. (1983). The measurement of user information satisfaction. *Communications of the ACM, 26,* 785–793.

John, G. (1984). An empirical investigation of some antecedents of opportunism in a marketing channel. *Journal of Marketing Research, 21,* 278–289.

John, G., & Martin, J. (1984). Effects of organizational structure of marketing planning on credibility and utilization of plan output. *Journal of Marketing Research, 21,* 170–183.

Johnson, D. W. (1975). Cooperativeness and social perspective taking. *Journal of Personality and Social Psychology, 31,* 241–244.

Johnson, D. W., & Lewicki, R. J. (1969). The initiation of superordinate goals. *Journal of Applied Behavioral Science, 5,* 9–24.

Johnson, J. P., & Dansby, H. B. (1995). Liability in private sector GIS. *Proceedings of the Conference on Law and Information Policy for Spatial Databases,* pp. 285–292.

Jordan, G. (2002). GIS for community forestry user groups in Nepal: Putting people before the technology. In W. J. Craig, T. M. Harris, & D. Weiner (Eds.), *Community participation and geographic information systems* (pp. 232–245). New York: Taylor & Francis.

Josephson, M. (2005). *Making sense of ethics*. Los Angeles: Josephson Institute of Ethics. Available at *www.josephsoninstitute.org/MED/MED-1makingsense. htm*

Kaplan, B., & Duchon, D. (1988). Combining qualitative and quantitative methods in information systems research. *MIS Quarterly, 12*(4), 571–586.

Keller, L. F. (1984). The political economy of public management. *Administration and Society, 15*, 455–474.

Kelley, H. H., & Stahelski, A. J. (1970). Social interaction basis of cooperators' and competitors' beliefs about others. *Journal of Personality and Social Psychology, 16*, 66–91.

Kelly, W. J. (2005, March 3). Leaving Bush behind: Companies make peace with Kyoto Protocol. *LA Weekly News*. Available at *www.laweekly.com/news/news/leaving_bush_behind/882*

Kim, K. (1998). *Using GIS technologies to empower community based organizations in Hawaii*. Presentation at the NCGIA Specialist Meeting on Empowerment, Marginalization, and Public Participation GIS, Santa Barbara, CA.

Kling, R. (1980). Social analyses of computing: Theoretical perspectives in recent empirical research. *Computing Surveys, 12*(1), 61–110.

Knox, P. L. (1988). Disappearing targets?: Poverty areas in central cities. *Journal of the American Planning Association, 54*, 501–508.

Kono, T. (1984). *Strategy and structure of Japanese enterprises*. Armonk, NY: M. E. Sharpe.

Kraemer, K. L., & Dutton, W. (1984). Survey research in the study of management information systems. In F. W. McFarlan (Ed.), *The information systems research challenge: Survey research methods* (pp. 3–58). Boston: Harvard Business School Press.

Kraemer, K. L., Dutton, W., & Northrop, A. (1981). *The management of information systems*. New York: Columbia University Press.

Kraemer, K. L., & Dutton, W. H. (1991). Survey research in the study of management information systems. In K. L. Kraemer (Ed.), *The information systems research challenge: Survey research methods* (Vol. 3, pp. 3–58). Boston: Harvard Business School.

Kraemer, K. L., King, J. L., Dunkle, D. E., & Lane, J. P. (1989). *Managing information systems: Change and control in organizational computing*. San Francisco: Jossey-Bass.

Kumar, K., & van Dissel, H. G. (1996). Sustainable collaboration: Managing conflict and cooperation in interorganizational systems. *MIS Quarterly, 20*(3), 279–300.

Laudon, K. C., & Laudon, J. P. (2001). *Essentials of management information systems*. Upper Saddle River, NJ: Prentice Hall.

Laughlin, P. R. (1978). Ability and group problem solving. *Journal of Research and Development in Education*, *12*, 114–120.

Lawrence, P. R., & Lorsch, J. W. (1967). Differentiation and integration in complex organizations. *Administrative Science Quarterly*, *11*, 1–47.

Lawrence, P. R., & Lorsch, J. W. (1969). *Organization and environment*. Homewood, IL: Irwin.

Layard, R., & Glaister, S. (Eds.). (1994). *Cost benefit analysis* (2nd ed.). New York: Cambridge University Press

Lee, A. S. (1989). A scientific methodology for MIS case studies. *MIS Quarterly*, *13*(1), 33–50

Leonard-Barton, D. (1987). Implementing structured software methodologies: A case of innovation in process technology. *Interfaces*, *7*(3), 6–17.

Levinsohn, A. G. (1989). A strategic planning based approach to the design of land information systems. *Proceedings of GIS/LIS '89*.

Lindahl, E. (1958). Just taxation: A positive solution. In R. Musgrave & J. Peacock (Eds.), *Classics in the economics of public finance*. London: Macmillan. (Original work published 1919)

Linden, D. S. (1991). Comments from the board: Ethics and the GIS industry—Are the terms contradictory? *GIS World*, *4*(4), 17.

Little, I. M. D., & Mirrlees, J. A. (1994). The costs and benefits of analysis: project appraisal and planning twenty years on. In R. Layard & S. Glaister (Eds.), *Cost benefit analysis* (2nd ed., pp. 199–234). New York: Cambridge University Press.

Longley, P. A., Goodchild, M., Maguire, D. J., & Rhind, D. W. (2001). *Geographic information systems and science*. New York: Wiley.

Lucas, H. C., Jr. (1975). Behavioral factors in system implementation. In R. L. Schultz & D. P. Slevin (Eds.), *Implementing operations research and management science* (pp. 203–216). New York: Elsevier.

Lucas, H. C., Jr., Ginzberg, M. J., & Schultz, R. L. (1990). *Information systems implementation: Testing a structural model (Computer-based information systems in organization)*. Norwood, NJ: Ablex.

Lucas, H. C., Jr., & Nielsen, N. R. (1980). The impact of the mode of information presentation on learning and performance. *Management Science*, *26*(10), 982–993.

MacNab, P. (2002). There must be a catch: Participatory GIS in a Newfoundland fishing community. In W. J. Craig, T. M. Harris, & D. Weiner (Eds.), *Community participation and geographic information systems* (pp. 173–191). New York: Taylor & Francis.

Maitra, J. B., Anderson, N., & Federal Geographic Data Committee (FGDC). (2005). Geospatial standards. Available at *fgdc.gov/publications/documents/standards/geospatial_standards_part1.html*

Majone, G., & Wildavsky, A. B. (1978). Implementation as evolution. *Policy Studies Review Annual*, *2*.

March, J. G. (1962). The business firm as a political coalition. *Journal of Politics*, *24*, 662–678.

March, J. G., & Simon, H. A. (1958). *Organizations*. New York: Wiley.

Mark, D. M. (2003). Geographic information science: Defining the field. In M. Duckham, M. Goodchild, & M. Worboys (Eds.), *Foundations of geographic information science* (pp. 1–15). New York: Taylor & Francis.

Markus, M. L. (1981). Implementation politics—Top management user involvement. *Systems, Objectives, Solutions, 2*, 203–215.

Markus, M. L. (1983). Power, politics and MIS implementation. *Communications of the ACM, 26*, 430–444.

Markus, M. L., & Bjorn-Andersen, N.(1987). Power-over users: Its exercise by system professionals. *Communications of the ACM, 30*(6), 498–504.

Markus, M. L., & Pfeffer, J. (1983). Power and the design and implementation of accounting and control systems. *Accounting, Organizations and Society, 8*, 205–218.

Masser, I., & Campbell, H. I. (1995). Information sharing: The effects of GIS on British local government. In H. I. Onsrud & G. Rushton (Eds.), *Sharing geographic information* (pp. 230–249). New Brunswick, NJ: Center for Urban Policy Research.

Matlack, C. (2006, October 23). Wayward Airbus. *BusinessWeek*, pp. 46–48.

Mayer, H. M. (1964). Politics and land use: The Indiana shoreline of Lake Michigan. *Annals of the Association of American Geographers, 54*.

McCann, J., & Galbraith, J. R. (1981). Interdepartmental relations. In P. C. Nystrom & W. H. Starbuck (Eds.), *Handbook of organizational design* (Vol. 2, pp. 60–84). New York: Oxford University Press.

McCann, J. E. (1983). Design guidelines for social problem-solving interventions. *Journal of Applied Behavioral Science, 19*(2), 177–192.

McDonnell, R., & Kemp, K. (1995). *International GIS dictionary*. New York: Wiley.

McGuire, M. (1974). Group segregation and optimal jurisdictions. *Journal of Political Economy, 80*, 112–132.

McNulty, K. (1994, January 9). The power of maps. *Chicago Tribune*, Section 5, p. 1.

Meredith, P. H. (1995). Distributed GIS: If its time is now, why is it resisted? In H. J. Onsrud & G. Rushton (Eds.), *Sharing geographic information* (pp. 7–21). New Brunswick, NJ: Center for Urban Policy Research.

Milton, P. (2007, February 20). Industry leaders fight greenhouse gases. *Forbes*. Available at *forbes.com/feeds/ap/2007/02/20/ap3446739.html*.

Milward, H. B. (1982). Interorganizational policy systems and research on public organizations. *Administration and Society, 13*, 457–478.

Mintzberg, H. (1979). *The structure of organizations*. Englewood Cliffs, NJ: Prentice Hall.

Mintzberg, H. (1983). *Power in and around organizations*. Englewood Cliffs, NJ: Prentice Hall.

Mintzberg, H. (1985). The organization as political arena. *Journal of Management Studies, 22*, 133–154.

Mitroff, I. I. (1975). On mutual understanding and the implementation problem:

A philosophical case study of the psychology of the Apollo moon scientists. In R. L. Schultz & D. P. Slevin (Eds.), *Implementing operations research and management science* (pp. 237–252). New York: Elsevier.

Moch, M., & Seashore, S. E. (1981). How norms affect behaviors in and of corporations. In P. C. Nystrom & W. H. Starbuck (Eds.), *Handbook of organizational design* (Vol. 1, pp. 534–565). New York: Oxford University Press.

Moenart, R. K., & Souder, W. E. (1990). An analysis of the use of extrafunctional information by R&D and marketing personnel: Review and model. *Journal of Product Innovation Management, 7,* 91–107.

Money, A., Tromp, D., & Wegner, T. (1988). The quantification of decision support benefits within the context of value analysis. *MIS Quarterly, 12*(2), 223–236.

Monmonier, M. (1991). *How to lie with maps.* Chicago: University of Chicago Press.

Monmonier, M. (1998). The three R's of GIS-based site selection: Representation, resistance, and ridicule. In D. R. F. Taylor (Ed.), *Policy issues in modern cartography* (pp. 233–248). Oxford, UK: Elsevier Science.

Monmonier, M. (2002). *Spying with maps.* Chicago: University of Chicago Press.

Montgomery, G., & Schuck, H. (1993). *GIS data conversion handbook.* Fort Collins, CO: GIS World.

Mooney, J. D. (1947). *The principles of organization.* New York: Harper.

Morris, M. (1996). *Planners salaries and employment trends.* Chicago: American Planning Association.

Mowshowitz, A. (1981). On approaches to the study of social issues in computing. *Communications of the ACM, 24*(3), 146–155.

MSNBC. (2004). FBI apologizes to lawyer held in Madrid bombings. Available at *www.msnbc.msn.com/id/5053007/*

Mumford, E., & Pettigrew, A. (1975). *Implementing strategic decisions.* London: Longman.

Murray, J.P. (2001). Finding the right level of IT expense. *Information Strategy: The Executive's Journal, 17*(2), 29–35.

Musgrave, R., & Peacock, A. T. (Eds.). (1958). *Classics in the theory of public finance.* New York: Macmillan.

National Research Council. (1993). *Toward a coordinated spatial data infrastructure.* Washington, DC: National Academy Press.

National Research Council, National Academy of Sciences, Committee on Review of GIS Research and Applications at HUD. (2003). *GIS for housing and urban development.* Washington, DC: National Academies Press.

Nedovic-Budic, Z., & Pinto, J. K. (2000). Information sharing in an interorganizational GIS environment. *Environment and Planning B: Planning and Design, 27,* 455–474.

Nedovic-Budic, Z., & Pinto, J. K. (2001). Organizational (soft) GIS interoperability: Lessons from the U.S. *International Journal of Applied Earth Observation and Geoinformation, 3*(3), 290–298.

Nedovic-Budic, Z., Pinto, J. K., & Warnecke, L. (2004). GIS database development and exchange: Interaction mechanisms and motivations. *Journal of the Urban and Regional Information System Association (URISA)*, *16*(1), 15–29.

Newman, W. H. (1988). *Role of departments in multi-level strategic management* (Working paper SC65). New York: Strategy Research Center, Columbia University.

NGDPF. (1993). *Present and emerging U.S. policies governing the development, evolution, and use of national spatial data infrastructure.* Report for the National Geo-Data Policy Forum. (Available from Federal Geographic Data Committee, U.S. Geological Survey, 590 National Center, Reston, VA 20192)

Niehoff, A. H. (1966). *A casebook of social change.* Chicago: Aldine.

North, R. C., Koch, H. E., Jr., & Zinnes, D. (1960). The integrative functions of conflict. *Journal of Conflict Resolution*, *4*(3), 355–374.

Oakland, W. H. (1972). Congestion, public goods and welfare. *Journal of Public Economics*, *1*(3), 339–357.

Obermeyer, N. J. (1990a). *Bureaucrats, clients and geography: The Bailly nuclear power plant battle in Northern Indiana* (University of Chicago Department of Geography Research Papers, No. 216). Chicago: University of Chicago Press.

Obermeyer, N. J. (1990b). Regional equity in turbulent times: The experience of the Regional Transportation Authority of Northeastern Illinois. *Applied Geography*, *10*, 147–161.

Obermeyer, N. J. (1992, April). *GIS: A new profession?* Presentation at the annual meeting of the Association of American Geographers, San Diego, CA.

Obermeyer, N. J. (1993). Certifying GIS professionals: Challenges and alternatives. *Journal of the Urban and Regional Information Systems Association*, *5*(1), 67–76.

Obermeyer, N. J. (1994). Spatial conflict in the information age. *URISA Annual Meeting Proceedings*, pp. 269–282.

Obermeyer, N. J. (1995a). GIS in society. (Chapter 19, Vol. 3: GIS Organization) In A. U. Frank (Ed.), *Geographic information systems: Materials for a postgraduate course: Vol. 3. GIS organization* (pp. 827–865). Vienna: COMMETT (Community Action Program in Education and Training for Technology of the European Union).

Obermeyer, N. J. (1995b). Reducing inter-organizational conflict to facilitate sharing geographic information. In H. J. Onsrud & G. Rushton (Eds.), *Sharing geographic information* (pp. 138–148). New Brunswick, NJ: Center for Urban Policy Research.

Obermeyer, N. J. (1998). Professional responsibility and ethics in the spatial sciences. In D. R. F. Taylor (Ed.), *Policy issues in modern cartography* (pp. 215–232). Oxford, UK: Elsevier.

Obermeyer, N. J. (1999). Measuring the benefits and costs of GIS. In P. A. Longley, M. F. Goodchild, D. J. Maguire, & D. W. Rhind (Eds.), *Geographic informa-*

tion systems: Management issues and applications (2nd ed., Vol. 2, pp. 601–610). New York: Wiley.

Obermeyer, N. J. (in press). Ethics and the social ideal in GIS. *Cartography and Geographic Information Science.*

Obermeyer, N. J., & Pinto, J. K. (1994). *Managing geographic information systems.* New York: Guilford Press.

Olson, M. (1971). *The logic of collective action: Public goods and the theory of groups.* Cambridge, MA: Harvard University Press.

Olson, M., & Zeckhauser, R. (1966). An economic theory of alliances. *Review of Economics and Statistics, 48,* 266–279.

Onsrud, H. J. (1989). Legal and liability issues in publicly accessible land information systems. *Proceedings of GIS/LIS '89, 1,* 295–300.

Onsrud, H. J. (1992). Evidence generated from GIS. *GIS Law, 1*(3), 1–9.

Onsrud, H. J. (1995). Identifying unethical conduct in the use of GIS. *Cartography and Geographic Information Science, 22*(1), 90–97.

Onsrud, H. J. (1999). Liability in the use of geographic information systems and geographic datasets. In P. Longley, D. Maguire, & D. Rhind (Eds.), *Geographical information systems: Vol. 2. Management issues and applications* (pp. 643–652). New York: Wiley.

Onsrud, H. J. (2005). GIS and the law. Online course available at the University of Maine.

Onsrud, H. J., Calkins, H. W., & Obermeyer, N. J. (Eds.). (1989). *Use and value of geographic information: Initiative Four specialist meeting summary report and proceedings* (Tech. Paper No. 89-7). Santa Barbara: University of California at Santa Barbara, National Center for Geographic Information and Analysis.

Onsrud, H. J., Johnson, J. P., & Lopez, X. (1994). Protecting personal privacy in using geographic information systems. *Photogrammetric Engineering and Remote Sensing, 60*(9), 1083–1095.

Onsrud, H. J., Johnson, J. P., & Winnecki, J. (1996). GIS dissemination policy: Two surveys and a suggested approach. *Journal of Urban and Regional Information Systems, 8*(2), 8–23. Also available online at *spatial.maine.edu/~onsrud/pubs/GIS_Dissemination/GIS_Diss_Policy.html*

Onsrud, H. J., & Lopez, X. (1998). Intellectual property rights in disseminating digital geographic data, products, and services: Conflicts and commonalities among European Union and United States approaches. In I. Masser & F. Salge (Eds.), *European geographic information infrastructures: Opportunities and pitfalls* (pp. 153–167). London: Taylor & Francis.

Onsrud, H. J., & Obermeyer, N. J. (1997). GIS education white paper. Available online at *http://157.182.168.41/ucgis_education.html*

Onsrud, H. J., & Pinto, J. K. (1991). Diffusion of geographic information innovations. *International Journal of Geographic Information Systems, 5,* 447–467.

Onsrud, H. J., & Pinto, J. K. (1992). *Correlating adoption factors and processes with*

GIS user satisfaction in U.S. local governments. Paper presented at NATO Advanced Research Workshop on Diffusion of Geographic Information Technologies, Sounion, Greece.

Onsrud, H. J., & Pinto, J. K. (1993). Evaluating correlates of GIS adoption success and the decision process of GIS acquisition. *Journal of the Urban and Regional Information Systems Association (URISA), 5*(1), 18–39.

Onsrud, H. J., Pinto, J. K., & Azad, B. (1992). Case study research methods for geographic information systems. *URISA Journal, 4*(1), 32–44.

Onsrud, H. J., & Rushton, G. (Eds.). (1995). *Sharing geographic information.* New Brunswick, NJ: Center for Urban Policy Research.

Orthner, H. F., Scherrer, J.-R., & Dahlen, R. (1994). Sharing and communicating health care information: Summary and recommendations. *International Journal of Bio-Medical Computing, 34,* 303–318.

Palast, G. (2003). *The best democracy money can buy.* New York: Plume

Papageorgiou, G. J. (1978). Spatial externalities: Parts I and II. *Annals of the Association of American Geographers, 68*(4), 465–492.

Parker, C., & Pascal, A. (2002). A voice that could not be ignored: Community GIS and gentrification battles in San Francisco. In W. J. Craig, T. M. Harris, & D. Weiner (Eds.), *Community participation and geographic information systems* (pp. 55–64). New York: Taylor & Francis.

Pavett, C. M., & Lau, A. W. (1983). Managerial work: The influence of hierarchical level and functional specialty. *Academy of Management Journal, 26,* 170–177.

Pearlson, K. E. (2001). *Managing and using information systems: A strategic approach.* New York: Wiley.

Perkins, H. (2005). Ethics for the GIS professional. *GISC-Eye, 1*(1), 7. Available at *www.gisci.org/GISC_Eye_vol1no1.pdf*

Perritt, H. H., Jr. (1995). Should local government sell local spatial databases through state monopolies? *Jurimetric Journal, 35,* 449–469.

Perritt, H. H., Jr. (1996). *Law and the information superhighway.* Chichester, UK: Wiley.

Perrow, C. (1967). A framework for the comparative analysis of organizations. *American Sociological Review, 70,* 686–704.

Perry, J. L., & Rainey, H. (1988). The public-private distinction in organization theory: A critique and research strategy. *Academy of Management Review, 13,* 182–201.

Peters, T. (1990). Get innovative or get dead! *California Management Review, 33,* 9–26.

Pettigrew, A. (1973). *The politics of organizational decision-making.* London: Tavistock.

Pettigrew, A. (1975). Toward a political theory of organizational intervention. *Human Relations, 28,* 191–208.

Pfeffer, J. (1981). *Power in organizations.* Marshfield, MA: Pitman.

Pfeffer, J. (1982). *Organizations and organization theory*. Boston: Pitman.

Pfeffer, J. (1992, Winter). Understanding power in organizations. *California Management Review*, pp. 29–50.

Pfeffer, J., & Salancik, G. (1978). Organization design: The case for a coalitional model of organizations. *Organization Dynamics, 6*, 15–29.

Pfeffer, J., & Salancik, G. R. (1978). *The external control of organizations*. New York: Harper & Row.

Pickles, J. (1993). Discourse on method and the history of discipline: Reflections on Dobson's 1983 *Automated Geography. Professional Geographer, 45*(4), 451–455.

Pickles, J. (1995). Representations in an electronic age: Geography, GIS, and democracy. In J. Pickles (Ed.), *Ground truth* (pp. 1–30). New York: Guilford Press.

Pinto, J. K. (1996). *Power and politics in project management*. Newtown Square, PA: Project Management Institute.

Pinto, J. K. (2000). Understanding the role of politics in successful project management. *International Journal of Project Management, 18*(2), 85–91.

Pinto, J. K., & Azad, B. (1994). The role of organizational politics in GIS implementation. *Journal of the Urban and Regional Information Systems Association (URISA), 6*(2), 35–61.

Pinto, J. K., & Onsrud, H. J. (1997). In search of the dependent variable: Toward synthesis in GIS implementation research. In M. Craglia & H. Couclelis (Eds.), *Geographic information research: Bridging the Atlantic* (pp. 129–145). London: Taylor & Francis.

Pinto, J. K., & Slevin, D. P. (1988). Project success: Definitions and measurement techniques. *Project Management Journal, 19*(1), 67–72.

Pinto, M. B., Pinto, J. K., & Prescott, J. E. (1993). Antecedents and consequences of project team cross-functional cooperation. *Management Science, 39*, 1281–1298.

Piore, M. J. (1979). Qualitative research techniques in economics. *Administrative Science Quarterly, 24*, 560–569.

Pondy, L. R. (1967). Organizational conflict: Concepts and models. *Administrative Science Quarterly, 12*(2), 296–320.

Porter, E. (1991a). GIS: Washington. *GIS World, 4*(4), 24, 27.

Porter, E. (1991b). Persian Gulf information restricted. *GIS World, 4*(4), 27.

Porter, M. E. (1979). How competitive forces shape strategy. *Harvard Business Review, 57*(2), 137–145.

Porter, M. E. (1985). *Competitive advantage*. New York: Free Press.

Pressman, J. L., & Wildavsky, A. B. (1973). *Implementation*. Berkeley: University of California Press.

Pugh, D. L. (1989). Professionalism in public administration: Problems, perspectives, and the role of ASPA. *Public Administration Review, 49*, 1–8.

Quinn, J. B. (1980). *Strategies for change*. Homewood, IL: Irwin.

Raghavan, S. A., & Chandf, D. R. (1989). Diffusing software-engineering methods. *IEEE Software, 15*(7), 81–90.

Raia, A. P. (1974). *Managing objectives*. Glenview, IL: Scott, Foresman.

Rasmussen, P. W. (1986). What should we do with the AICP exam? *Journal of the American Planning Association, 52*, 7–8.

Raymond, L. (1987). Validating and applying user satisfaction as a measure of MIS success in small organizations. *Information and Management, 12*, 173–179.

Reich, B. H., & Huff, S. L. (1991). Customer oriented strategic systems. *Journal of Strategic Information Systems, 1*(1), 29–37.

Rhind, D. (1996). Data access, charging and copyright and their implications for geographical information systems. *International Journal of Geographical Information Systems, 6*(1), 13–30.

Rhind, D. W. (1999). National and international geospatial data policies. In P. A. Longley, M. F. Goodchild, D. J. Maguire, & D. W. Rhind (Eds.), *Geographic information systems: Vol. 2. Management issues and applications* (2nd ed., pp. 767–787). New York: Wiley.

Rice, R. E., & Rogers, E. M. (1980). Re-invention in the innovation process. *Knowledge: Creation, Implementation, Utilization, 1*, 499–514.

Robbins, S. P. (1992). *Essentials of organizational behavior* (4th ed.). Englewood Cliffs, NJ: Prentice Hall.

Roberts, J. L. (2003, August 4). Big media's big headache. *Newsweek*, p. 44.

Robey, D. (1984). Conflict models for implementation research. In R. L. Schultz & M. J. Ginsberg (Eds.), *Management science implementation*. Greenwich, CT: JAI Press.

Robey, D., & Farrow, D. (1982). User involvement in information systems development: A conflict model and empirical test. *Management Science, 28*(1), 73–85.

Robey, D., Farrow, D., & Frantz, C. (1989). Group process and conflict in system development. *Management Science, 35*(10), 1172–1191.

Robey, D., & Markus, M. L (1984). Rituals in information systems design. *MIS Quarterly, 8*(1), 5–15.

Robey, D., Smith, L., & Vijayasarathy, L. (1993). Perceptions of conflict and success in information systems development projects. *Journal of Management Information Systems, 10*(1), 123–139.

Rogers, E. M. (1962). *Implementation of innovations*. New York: Free Press.

Rogers, E. M. (1983). *Implementation of innovations* (3rd ed.). New York: Free Press.

Roitman, H. (1988). Public records laws: A proposed model for changes. *Proceedings of the 26th Annual Urban and Regional Information Systems Association, 4*, 338–347.

Ruekert, R. W., & Churchill, G. A., Jr. (1984). Reliability and validity of alternative measures of channel member satisfaction. *Journal of Marketing Research, 21*, 226–233.

Ruekert, R. W., & Walker, O. C. (1987a). Interactions between marketing and R&D departments in implementing difference business strategies. *Strategic Management Journal, 8,* 233–248.

Ruekert, R. W., & Walker, O. C. (1987b). Marketing's interaction with other functional units: Conceptual framework and empirical evidence. *Journal of Marketing, 51,* 1–19.

Rundstrom, R. A. (1995). GIS, indigenous peoples, and epistemological diversity. *Cartography and Geographic Information Systems, 22*(1), 45–57.

Salge, F. (1999). National and international data standards. In P. A. Longley, M. F. Goodchild, D. J. Maguire, & D. W. Rhind (Eds.), *Geographical information systems: Vol. 1. Principles and technical issues* (2nd ed., pp. 693–706). New York: Wiley.

Samuelson, P. A. (1954). The pure theory of public expenditures. *Review of Economics and Statistics, 36*(4), 387–389.

Sawicki, D., & Burke, P. (2002). The impacts of GIS use for neighbourhood revitalization in Minneapolis. In W. J. Craig, T. M. Harris, & D. Weiner (Eds.), *Community participation and geographic information systems* (pp. 89–100). New York: Taylor & Francis.

Sawicki, D. S., & Peterman, D. (2002). Surveying the extent of PPGIS practice in the United States. In W. J. Craig, T. M. Harris, & D. Weiner (Eds.), *Community participation and geographic information systems* (pp. 17–36). New York: Taylor & Francis.

Schermerhorn, J. R. (1975). Determinants of inter-organizational cooperation. *Academy of Management Journal, 18,* 846–856.

Schermerhorn, J. R., Jr. (1989). *Managing for productivity.* New York: Wiley.

Schmidt, W. H., & Tannenbaum, R. (1960). The management of scientific manpower. *Management Science, 14,* B473–B489.

Schultz, R. L., Ginzberg, M. J., & Lucas, H. C., Jr. (1983). *A structural model of implementation.* Working paper, University of Texas at Dallas.

Schultz, R. L., & Slevin, D. P. (1975). Implementation and management innovation. In R. L. Schultz & D. P. Slevin (Eds.), *Implementing operations research/ management science* (pp. 3–20). New York: Elsevier.

Schultz, R. L., & Slevin, D. P. (1979). Introduction: The implementation problem. In R. Doktor, R. L. Schultz, & D. P. Slevin (Eds.), *The implementation of management science* (pp. 1–15). New York: North-Holland.

Schultz, R. L., Slevin, D. P., & Pinto, J. K. (1987). Strategy and tactics in a process model of project implementation. *Interfaces, 17*(3), 34–46.

Schuurman, N. (2004). *GIS: A short introduction.* Oxford, UK: Blackwell.

Scott, B. W. (1963). *Some aspects of long-range planning in American corporations with special attention to strategic planning.* Unpublished doctoral dissertation, Harvard University, Cambridge, MA.

Sen, A. K. (1994). Shadow prices and markets: Feasibility constraints. In R. Layard

& S. Glaister (Eds.), *Cost benefit analysis* (2nd ed., pp. 59–99). New York: Cambridge University Press.

Sense, A. J. (2003). A model of the politics of project leader learning. *International Journal of Project Management, 21*(2), 107–114.

Shapiro, B. P. (1977). Can marketing and manufacturing coexist? *Harvard Business Review, 55*, 104–114.

Shapiro, E. (1993, December 16). Cigarette maker and Time aim ads at smokers. *Wall Street Journal*, pp. B1, B7.

Sheppard, E. (1993). Automated geography: What kind of geography for what kind of society. *Professional Geographer, 45*(4), 457–460.

Sheppard, E. (1995). GIS and society: Toward a research agenda. *Cartography and Geographic Information Systems, 22*(1), 5–16.

Sherif, M. (1962). Intergroup relations and leadership: Introductory statement. In M. Sherif (Ed.), *Intergroup relations and leadership* (pp. 3–21). New York: Wiley.

Sherif, M., & Sherif, C. W. (1969). *Social psychology*. New York: Harper & Row.

Sieber, R. E. (2002). Geographic information systems in the environmental movement. In W. J. Craig, T. M. Harris, & D. Weiner (Eds.), *Community participation and geographic information systems* (pp. 153–172). New York: Taylor & Francis.

Simon, H. (1964). On the concept of organizational goals. *Administrative Science Quarterly, 9*, 1–22.

Simon, H. A. (1976). *Administrative behavior: A study of decision-making processes in administrative organizations*. New York: Free Press. (Original published 1945)

Slevin, D. P. (1989). *The whole manager: How to increase your professional and personal effectiveness*. New York: American Management Association.

Slevin, D. P., & Pinto, J. K. (1987). Balancing strategy and tactics in project implementation. *Sloan Management Review, 29*, 33–41.

Smith, D. A., & Tomlinson, R. F. (1992). Assessing the costs and benefits of geographical information systems: Methodological and implementation issues. *International Journal of Geographical Information Systems, 6*(3), 247–256.

Smith, D. K., & Alexander, R. (1988). *Fumbling the future: How Xerox invented, then ignored, the first personal computer*. New York: Morrow.

Smith, N. (1992). History and philosophy of geography: Real wars, theory wars. *Progress in Human Geography, 16*(2), 257–271.

Sommers, R. (1990). Organizational structures in GIS development. In *Proceedings from the 1990 Annual Conference of the Urban and Regional Information Systems Association, 5*.

Souder, W. E. (1981). Disharmony between R&D and marketing. *Industrial Marketing Management, 10*, 67–73.

Souder, W. E. (1988). Managing relations between R&D and marketing in new product development projects. *Journal of Product Innovation Management, 5*, 6–19.

Sperling, J. (1995). Development and maintenance of the TIGER database: Experiences in spatial data sharing at the U.S. Bureau of the Census. In H. J. Onsrud & G. Rushton (Eds.), *Sharing Geographic Information* (pp. 377–396). New Brunswick, NJ: Center for Urban Policy Research.

Srinivasan, A., & Davis, J. G. (1987). A reassessment of implementation process models. *Interfaces, 17*(3), 64–71.

Stern, L. W., & Heskett, J. L. (1968). Conflict management in inter-organizational relations: A conceptual framework. In L W. Stern (Ed.), *Distribution channels: Behavioral dimensions* (pp. 288–305). Boston: Houghton Mifflin.

Stern, L. W., Sternthal, B., & Craig, C. S. (1973). Managing conflict in distribution channels: A laboratory study. *Journal of Marketing Research, 10,* 169–179.

Stevens, J. M., Wartick, S., & Bagby, J. (1988). *Business–government relations and interdependence: A managerial and analytical perspective.* New York: Praeger.

Stiglitz, J. E. (1994). Discount rates: The rate of discount for benefit–cost analysis and the theory of the second best. In R. Layard & S. Glaister (Eds.), *Cost benefit analysis* (2nd ed., pp. 116–159). New York: Cambridge University Press.

Stough, R. R., & Whittington, D. (1985). Multijurisdictional waterfront land use modeling. *Coastal Zone Management Journal, 13,* 151–175.

Tarter, B. (1992). Information liability: New interpretations for the electronic age. *Computer/Law Journal, 11,* 481–553.

Taupier, R. P. (1995). Comments on the economics of geographic information and data access in the Commonwealth of Massachusetts. In H. J. Onsrud & G. Rushton (Eds.), *Sharing geographic information* (pp. 277–291). New Brunswick, NJ: Center for Urban Policy Research.

Taylor, F. W. (1911). *Principles of management.* New York: Harper.

Thomas, K. (1976). Conflict and conflict management. In M. D. Dunnette (Ed.), *Handbook of industrial and organizational psychology* (pp. 889–935). Chicago: Rand McNally.

Thompson, A. A., Jr., & Strickland, A. J., III. (1987). *Strategic management: Concepts and cases* (12th ed.). Burr Ridge, IL: McGraw-Hill.

Thompson, J. (1967). *Organizations in action.* New York: McGraw-Hill.

Tiebout, C. (1956). A pure theory of local government expenditure. *Journal of Political Economy, 64,* 416–424.

Tobler, W. R. (1970). A computer movie simulating urban growth in the Detroit region. *Economic Geography, 46,* 234–240.

Tobler, W. R. (1976). Analytical cartography. *American Cartographer, 3*(1), 21–31.

Tomlinson, R. (2005). *Thinking about GIS: Revised and updated edition.* Redlands, CA: ESRI Press.

Trist, E. (1977). Collaboration theory and organizations. *Journal of Applied Behavioral Science, 13,* 268–278.

Tulloch, D. L. (2002). Environmental NGOs and community access to technology as a force for change. In W. J. Craig, T. M. Harris, & D. Weiner (Eds.), *Com-*

munity participation and geographic information systems (pp. 192–204). New York: Taylor & Francis.

Tushman, M. L. (1977). A political approach to organizations: A review and rationale. *Academy of Management Review, 2*, 206–216.

URISA. (2002). A GIS code of ethics. See URISA Website online at *www.urisa.org/ethics/code_of_ethics.htm*.

U.S. Bureau of the Budget. (1947). United States national map accuracy standards. Available at *rockyweb.cr.usgs.gov/nmpstds/acrodocs/nmas/NMAS647.PDF*

U.S. Census Bureau. (2007). Centers of population for Census 2000. Available at *www.census.gov/geo/www/cenpop/cntpop2k.html*

U.S. Geological Survey. (1999). Map accuracy standards: Fact sheet FS-171-99 (November 1999). Available at *erg.usgs.gov/isb/pubs/factsheets/fs17199. html*

Van de Ven, A. H. (1976). On the nature, formation, and maintenance of relations among organizations. *Academy of Management Review, 1*(4, 23–44.

Van de Ven, A. H., Delbecq, A. L., & Koenig, R., Jr. (1976). Determinants of coordination modes within organizations. *American Sociological Review, 41*, 322–338.

Vanlommel, E., & DeBrabander, B. (1975). The organization of electronic data processing (EDP) activities and computer use. *Journal of Business, 48*(4), 391–410.

Ventura, S. J. (1995). Overarching bodies for coordinating geographic data sharing at three levels of government. In H. J. Onsrud & G. Rushton (Eds.), *Sharing geographic information* (pp. 172–192). New Brunswick, NJ: Center for Urban Policy Research.

Veregin, H. (1999). Data quality parameters. In P. A. Longley, M. F. Goodchild, D. J. Maguire, & D. W. Rhind (Eds.), *Geographical information systems: Vol. 1. Principles and technical issues* (2nd ed., pp. 177–189). New York: Wiley.

Walker, D. H., Leitch, A. M., DeLai, R., Cottrell, A., Johnson, K. L., & Pullar, D. (2002). A community-based and collaborative GIS joint venture in rural Australia. In W. J. Craig, T. M. Harris, & D. Weiner (Eds.), *Community participation and geographic information systems* (pp. 137–152). New York: Taylor & Francis.

Wall, J. A., Jr. (1985). *Negotiation theory and practice.* New York: Scott, Foresman.

Walsh, K., Hinings, B., Greenwood, R., & Stewart, R. (1981). Power and advantage in organizations. *Organization Studies, 2*(2), 131–152.

Walton, R. E., & Dutton, J. M. (1969). The management of interdepartmental conflict: Model and review. *Administrative Science Quarterly, 14*, 73–84.

Wang, M., & Stough, R. R. (1986). Cognitive analysis of land-use decision-making. In *Modeling and simulation: Vol. 17, Part 1. Geography–Regional sciences, economics* (Proceedings of the 17th annual Pittsburgh conference, pp. 107–112).

Waters, N. (1992). Edge notes: Peace on Earth: Using GIS for military purposes. *GIS World, 5*(10), 76.

Weber, M. (1946). Bureaucracy. In H. H. Gerth & C. Wright Mills (Eds. & Trans.), *From Max Weber: Essays in sociology* (pp. 196–244). New York: Oxford University Press.

Weber, M. (1968a). *Economy and society* (G. Roth & C. Wittich, Eds.). Berkeley: University of California Press.

Weber, M. (1968b). Sociological categories of economic action. In G. Roth & C. Wittich, Eds. & Trans.), *Economy and society* (Vol. 1, pp. 63–210). Berkeley: University of California Press.

Weber, M. (1978). *Economy and society*. Berkeley: University of California Press.

Weick, K. (1978). Educational organizations as loosely coupled systems. *Administrative Science Quarterly*, *23*(3), 541–552.

Weiner, D., Warner, T., Harris, T., & Levin, R. (1995). Apartheid representations in a digital landscape: GIS, remote sensing and local knowledge in Kiepersol, South Africa. *Cartography and Geographic Information Systems*, *22*(1), 30–44.

Weiss, H. L. (1983). Why business and government exchange executives. In J. L. Perry & K. L. Kraemer (Eds.), *Public management*. Mountain View, CA: Mayfield.

Wellar, B. (1988a). A framework for research on research in the information technology–local government field. *Proceedings of the 26th Annual Conference of the Urban and Regional Information Systems Association*, *4*, 379–386.

Wellar, B. (1988b). Institutional maxims and conditions for needs-sensitive information systems and services in local governments. *Proceedings of the 26th Annual Conference of the Urban and Regional Information Systems Association*, *4*, 371–378.

Wentworth, M. J. (1989). Implementation of a GIS project in a local government environment: The long and winding road. *Proceedings of the 27th Annual Urban and Regional Information Systems Association*, *2*, 198–209.

Whiteman, J. (1983). Deconstructing the Tiebout hypothesis. *Environment and Planning D: Society and Space*, *1*(3), 339–354.

Wigand, R. T. (1988). Integrated telecommunications, networking, and distributed data processing. In J. Rabin & E. M. Jackowski (Eds.), *Handbook of Information Resource Management* (pp. 293–321). New York: Dekker.

Wiggins, L. L., & French, S. P. (1991). *GIS: Assessing your needs and choosing a system* (Planning Advisory Service Report 433). Chicago: American Planning Association.

Wilcox, D. L. (1990). Concerning "The economic evaluation of implementing a GIS." *International Journal of Geographical Information Systems*, *4*(2), 203–210.

Williams, A. (1966). The optimal provision of a public good in a system of local government. *Journal of Political Economy*, *74*, 18–33.

Williams, F., Rice, R. E., & Rogers, E. M. (1988). *Research methods and the new media*. New York: Free Press.

Williamson, O. E. (1975). *Markets and hierarchies: Analysis and antitrust implications*. New York: Free Press.

Wilson, D. C. (1982). Electricity and resistance: A case study of innovation and politics. *Organization Studies, 3*(2), 119–140.

World Intellectual Property Organization. (2006). See website online at *www.wipo.int/about-wipo/en/gib.htm.*

Yapa, L. (1998). Why GIS needs postmodern social theory, and vice versa. In D. R. F. Taylor (Ed.), *Policy issues in modern cartography* (pp. 249–270). Oxford, UK: Elsevier Science.

Yates, D. (1985). *The politics of management: Exploring the inner workings of public and private organizations.* San Francisco: Jossey-Bass.

Zaltman, G., & Duncan, R. (1977). *Strategies for planned change.* New York: Wiley.

Zaltman, G., & Moorman, C. (1989). The management and use of advertising research. *Journal of Advertising Research, 28,* 11–18.

Zelinsky, W. (1980). North America's vernacular regions. *Annals of the Association of American Geographers, 70,* 1–16.

Zerbe, R. O., Jr., & Dively, D. D. (1994). *Benefit–cost analysis in theory and practice.* New York: HarperCollins.

Zey, M. (1991). Criticisms of the rational choice models. In M. Zey (Ed.), *Decision making: Alternatives to Rational Choice Models* (pp. 9–31). Newbury Park, CA: Sage.

Zwart, P. (1991, August). Some indicators to measure the impact of land information systems in decision making. *Proceedings of the 1991 Conference of the Urban and Regional Information Systems Association.*

Index

Page numbers followed by *f* indicate figure, *t* indicate table

346

About the Authors

Nancy J. Obermeyer, PhD, is Associate Professor of Geography at Indiana State University. Her research interests include GIS implementation issues, public-participation GIS, professionalism, and ethics. Dr. Obermeyer began her professional life in several Illinois state agencies, working as an analyst in the Office of Planning, an energy planner in the Department of Energy and Natural Resources, and a project manager in the Department of Transportation. She was a member of the founding Board of Directors of the GIS Certification Institution (GISCI), and currently serves on GISCI's Ethics Committee.

Jeffrey K. Pinto, PhD, is Professor of Management at Penn State Erie, The Behrend College. His research interests include project management, information system implementation, power and political processes in organizations, and the diffusion of innovations. Dr. Pinto holds the Andrew Morrow and Elizabeth Lee Black Chair in Management of Technology, has received the Distinguished Contribution Award from the Project Management Institute and the Behrend Council of Fellows Research Award, and has consulting experience with a number of major organizations.